I0504513

Secrets Of
Anti-Aging
Larry White

Chapter 1
6 Fantastic Clothing Tips That'll Make You Look 5 Years Younger

Many of us would never turn down the chance to look a little younger — maybe by around five years. Well, I am here to prove to you that this isn't just fantasy. Here are six practical ways you can adjust your appearance to look like you've lost a few years.

Be bold
It seems that our obsession with looking 'presentable' sometimes leads us to be a little boring and old-fashioned with our style. Whatever your age, you can always be a little bold in the accessories you wear. All you have to do is add one exciting and unexpected element to your outfit. This might be a bright scarf, skirt, or blouse with an unusual print.

Introduce a little dynamism
A simple striped top can make you look more cheerful and vigorous. In this case, your preference should be for one with narrow horizontal stripes, as wider stripes will make you look larger. Clothes with spotted patterns will also have the same positive effect on your appearance.

Emphasize your unique style
Add accessories to let the real you shine through. One large necklace, a large pair of earrings, or a bracelet will stand out especially well. In this case, it's best to forgo wearing any more, or otherwise wear only very small ones that are almost unnoticeable.

Illuminate your face

The lighter your face looks, the younger and more positive you appear. An effective way to brighten your face is to wear something with a white collar. A polo shirt works better than a blouse here.

Wear your jeans

Jeans are the perfect choice for emphasizing your figure and hiding your imperfections, and they should be an essential element of everyone's wardrobe. They emphasize youth and can thus do a great deal to knock a couple of years off your appearance.

A question of color

I was getting my hair done the other day and my stylist and I started discussing hair colour and how it works with our skin tone. In recent years I've stopped dying my hair a blue-black and softened it to a very dark brown, one shade below black. While I still love the intensity of my other colour, it was starting to look too harsh. Of course, I still wear black clothing all the time, but I've definitely noticed how much brighter my complexion looks when I add a little bit of colour near my face. It was no different when I bought my latest hat (a fisherman's cap in case you wanted to know), but rather than black I opted for a dark navy because it had a bigger contrast with my hair and gave my look more dimension.

It's not just about black though. I know you thought it was just that simple, but hear me out because it's not complicated. It is, however, about making small adjustments to the colours you choose so that you can highlight your best features.

This isn't about wearing or not wearing whatever you want because you are welcome to do whatever you like, but it's more about not ageing yourself. I'm of the firm belief that no one wants to intentionally look older than they are.

BEST COLOURS TO KEEP YOU LOOKING YOUNGER

Black. Okay, let's just get this one out of the way first. Not all black shades are created equal. It can be faded or it can lean toward a grey, graphite or it can be a deep true black, the latter being the harshest of the three. Where a deep black can highlight your fine lines or the circles under your eyes try a softer shade which will still be black but not as harsh.

If you absolutely have to wear it near your face then soften your makeup and add a flattering blush to give your face more colour.

Turquoise. This is a great colour, that goes with every skin tone and every age. One small tip, if you aren't sure about the best shade for you, simply match it to the veins in your arms. My skin is quite pale and my veins are are more green so I simply reach for a greener shade of turquoise. If yours are bluer then that's the shade you should be looking for. For those with a more medium skin tone, any shade will do.

Neutrals. Like black, which gets its own category, other neutrals can either brighten or dull your complexion. You might look fabulous in crisp white or it might start washing your out and in that case you will need to reach for a cream or beige. On that note, a neutral grey will look great on pretty much everyone.

Red. You'll always be safe when reaching for a primary red. Nothing more needs to be said, find a true red and wear it with pride.

Pink. This is the colour of youth, but if you have very pale skin then a soft pastel shade might not be the best choice. Instead, brighten the colour by reaching for a red-violet tone. A jewel toned magenta, raspberry or fuschia will all be great.

Lavender-Blue. This probably isn't my favourite colour to wear, but if you like soft blue-violet shades then a periwinkle will also make you look more youthful.

Clothing in colors such as lemon yellow, mint green, and coral red can work wonders for making you look younger. This might be a dress, skirt, or blouse.
In this case, it's better to go for simple and unelaborate designs.

A few more useful tips
1. Select things made from natural materials with an even surface. Crumpled as well as overly tight items of clothing can emphasize wrinkles, especially if we're talking about blouses, jackets, scarves, and other things which come close to the face. Things made from tweed and other similar materials can also make you look older.
2. Stay away from matching items of clothing, such as a top in the exact same style as a skirt or pants. The only place this kind of style works is in the office. In all other situations, choose items of clothing that are similar but not identical in tone/material.
3. Black, gray, and beige clothing shouldn't dominate your appearance. Wearing one item of any of these colors is enough at any one time.

Chapter 2

Full Lips Make You Look Younger

In today's society, there's nothing that says beauty quite like a pair of plump, pouty lips. And, while we all know lip enhancement can make a woman look *better*, patients are always curious if it can really make a woman look *younger*. Yes it can, suggests beauty and fashion resource, *Redbook*.

The magazine recently published an interesting article entitled, "Is Your Face Older Than You Are?" Besides the obvious issues that make a woman look older – wrinkles, large pores, veined hands – the article also offered one factor that may surprise you: a woman's lips. According to *Redbook*, "A study by David Gunn, a research scientist at Unilever (makers of Dove, Vaseline, and more), found that women who look young for their age have fuller lips." So, just how much younger can a woman with full lips expect to look? While everyone is different, according to *Redbook*, full lips can take away as many as five years off a woman's appearance – a number that's hard to ignore!

How to Get Full Lips

If the Kardashian/Jenner family has taught us anything, it's that a heavy contour and an over-lined lip is almost as transformative as an actual surgical procedure – without the cost and recovery time. So if you're trying to get that thicc pout like our lord and savior Kylie Jenner, just steal the tricks she used *before* she started getting lip fillers on the reg. Pre-injections (think, 2014 Tumblr Kylie), she used to hack her way to fuller lips and the outcome looked so realistic, people seriously thought she got injections. Her secret? Using some pretty simple makeup techniques and affordable drugstore treatments that pretty much anyone can do. Basically, you just have to over-line, exfoliate, and use your concealer like a pro – then, top it all off with a little swipe of lip-plumping gloss. Easy!

So before you start Googling local aestheticians, check out all the non-permanent hacks that will level-up your lips in seconds.

1. Exfoliate

This is step number one and it's *so* important. Before you try any other hacks, exfoliate your lips first. Flaky lips reflect less light, which means, a dry pucker could be making your lips look smaller than they actually are. To get rid of flakes, brush them lightly with a toothbrush. This also boosts circulation, giving you a rosy tint sans makeup.

If you want to get professional with your exfoliation game, get yourself a hydrating sugar lip scrub that *really* smooths. Yes, it's another step to add to your skincare routine, but trust me, it's soooo worth it.

2. Hydrate, hydrate, hydrate

Have you ever noticed that when you forget your Hydro Flask at home, your lips look really freaking small by the end of the day? When your body is dehydrated, your lips literally shrivel up and disappear. Make sure to drink lots of water – which, btw, is just a good life choice in general – to bring out your most luscious lippies naturally (and for free).

3. Over-line your lips

Before Kylie Jenner surgically enhanced her lips, she used to use this trick to make them appear bigger. Now, you can't just swipe lipstick all over your face – you'll end up looking like Miranda Sings. The trick is to put on concealer *before* your lip color to blur the line between your lips and your face.

Even though this makeup trick is a beauty vlogger favorite, it's really easy to do yourself. Apply your go-to concealer to your lips, then trace slightly outside your natural lip line using a lip liner. From there, just fill in your lips with the matching lipstick or just use the liner for your whole pout.

4. Highlight with gloss

This two-second trick will help draw attention to your lips. Grab your favorite clear lip gloss or a pick shade that matches your lipstick color, then apply a dab of the gloss onto the middle of both of your lips. The shimmery product picks up light, making your pout look fuller.

5. Try a plumping treatment

The Dr. Dennis Gross Hyaluronic Marine Collagen Lip Cushion is a plumping lip treatment that delivers definition to the lip line and boosts hydration to enhance volume. There are tons of great plumping balms, glosses, and treatments out there, but this one is my personal favorite. It *seriously* works like a dream. After one application, I noticed a much fuller lip look — no needles necessary!

6. Double up on nude lip colors

Looking for bigger lips? What you really need is dimension, sis. Layering color amps up your pout easily. Here's how: Apply one shade all over your mouth, then top it with a slightly lighter color (or even a white lip crayon) in the middle of your lips. Along with making your lips look bigger, it'll help even out your lips. If you have a fuller top lip and a smaller bottom lip (or vise versa), try only applying the lighter shade to the smallest lip.

7. Line with concealer *after* lipstick

Ever wonder how influencers get that FaceTune-perfect lipstick line? Well, partly from FaceTune, but it's mostly thanks to this concealer hack. Yes, you should be applying concealer before *and* after your lipstick. Use a skinny brush and a teeny tiny bit of cover-up to line the outside of your already over-lined lips. You won't believe how much your lipstick will pop.

8. Don't skip the lip balm

I'll admit, lip balm isn't as fun to shop for as a new eyeshadow palette, but it's an absolutely essential part of your beauty routine. Your lips are delicate, so they need to be hydrated inside and out. Drink your water (see point 2) and then swipe on a lip balm that will soften your lips. Healthy lips will reach their highest, fullest potential.

Chapter 3
Want To Look 24 Percent Younger? Wear Sunscreen

With the summer months approaching, we're thinking about sunscreen — and we have a whole new reason to stock up on SPF. A new **study** published in the *Annals of Internal Medicine* says that sunscreen decreases skin aging by 24 percent.

Dermatologists have been warning people for years that unprotected UV exposure accelerates aging, but this is the first major study to prove that wearing sunscreen decreases wrinkles.

In the study, 903 participants, all younger than 55 and living near the sunny coast of Australia, were given SPF 15 sunscreen. Half were told to apply it daily while also reapplying after swimming, heavy sweating or just spending several hours in the sun. The other half were told to apply it as they normally would.

To measure the sunscreen's effects, researchers took a silicone impression of the back of subjects' hands at the beginning and again four years later at the end of the study to look for skin coarseness and patterns of lines.

Those who were instructed to wear sunscreen daily showed 24 percent less skin aging than those who didn't. And if that's just over four years, imagine how much younger a regular sunscreen wearer would look after, say, 10 or 20.

The moral of the story? Sunscreen doesn't just help you look younger, longer — wearing it helps you look considerably less wrinkled. And don't forget to apply it on the backs of your hands, which along with the face, show signs of aging before other areas.

We all know that wearing sunscreen each and every day is imperative for shielding the skin from the UV damage that leads to visible signs of aging and skin cancer. But according to a new study published in the Dermatological Surgery journal, sunscreen can actually improve existing signs of aging as well.

This study followed 32 people who applied broad-spectrum SPF 30 for 52 weeks. As early as 12 weeks into the study, researchers saw improvement in skin tone, discoloration, crow's feet, skin clarity, skin texture and overall sun damage. Study participants were also told not to use any over-the-counter or prescription anti-aging or anti-acne products over the course of the trial, and to avoid excessive sun exposure.

I am not at all surprised by these findings, and I'm glad this has finally been proven. This demonstrates the skin's ability to rejuvenate itself by continuing to produce collagen when it is shielded from environmental damage. When the factors that contribute to collagen breakdown are eliminated, the skin becomes thicker and smoother. Famed dermatologist, Dr. Albert Kligman, used to tell the story of a coma patient in a windowless room whose skin improved year after year. Although he never proved this was due to a lack of sun exposure, he was definitely onto something!

These findings reinforce what dermatologists have been saying for years: Sunscreen is one of the most effective anti-aging products available today. I know this will make it easier to convince all my patients that they should be wearing sunscreen daily (in addition to using an antioxidant serum) not only to prevent visible signs of aging, but to improve their appearance as well.

However, a vast majority of people don't use enough sunscreen, or apply it incorrectly. To reap the age-reversing benefits of SPF, be sure to follow these tips…

- Use ½ teaspoon of sunscreen on your face, and another ½ teaspoon for your neck and chest.
- Do not rely on the SPF in your foundation or powder for sun protection.
- Use an SPF of at least 15 every day, regardless of whether you'll be spending time outdoors.
- Use an SPF of at least 45 if you plan on being outside for more than 15 minutes, and be sure to reapply every hour.
- If you're expecting more than one hour of direct sun, first apply a chemical-based sunscreen (with ingredients such as Mexoryl or avobenzone) and follow with a second layer of a physical sunscreen with zinc oxide.
- For maximum skin protection, take Heliocare, which is an oral antioxidant supplement.
- No sunscreen is perfect, so wear a hat, clothing with UPF (UV protection) and stay in the shade when possible.

- A regular tee shirt only has an SPF of 5. Hold fabric up to the light, and the amount that shines through gives you an idea of its sun-protection level.
- My chemical sunscreen recommendations are Neutrogena Ultra Sheer Dry-Touch, EltaMD UV Clear Broad-Spectrum SPF 46 and La Roche-Posay Anthelios.
- My physical sunscreen recommendations are Obagi Sun Shield Tint Broad Spectrum SPF 50, PCA Skin Weightless Protection Broad Spectrum SPF 45, EltaMD UV Physical Broad-Spectrum SPF 41 and Blue Lizard Sensitive.

Chapter 4
Want to Look Younger? Your Eyebrows May Be the Key
Maybe there's some science behind the dramatic eyebrow trend after all.

A new study finds that facial features, like lips and eyebrows, tend to stand out less as people get older. Because of that, the authors say, people perceive faces with more contrast as younger.

In the study, published in the journal Frontiers in Psychology, researchers analyzed photographs of 763 makeup-free women with various skin tones between ages 20 and 80. A computer program analyzed the photos for facial contrast: a measure of how much the eyes, lips and eyebrows stand out due to differences in color, lightness or darkness with the surrounding skin.

Younger women had more facial contrast, and older women had less. Contrast especially decreased in areas around the mouth and eyebrows as women got older.

Next, the researchers Photoshopped some of the pictures, creating two nearly identical versions of each face with varying levels of contrast. They showed these photographs to volunteers and asked them to choose the younger-looking face. Almost 80% of the time, people said the high-contrast face appeared younger than the low-contrast one.

The findings were similar across a variety of ethnicities, suggesting that facial contrast—like wrinkles and changes in skin pigmentation—is truly a "cross-cultural cue" for perceiving how old a person is, the authors wrote in their paper.

Anyone who's ever filtered a selfie on Instagram won't be surprised by this effect of contrast. But the findings may also help explain why people often use makeup to look younger.

The study didn't involve makeup, so the authors cannot say for sure that darkening features cosmetically would have the same anti-aging results as demonstrated in the study. "But the way we manipulated features in the photos was very similar to what you'd do with makeup, and I would be surprised if you couldn't get similar effects," says co-author Richard Russell, associate professor of psychology at Gettysburg College. "We know that lips get less red with age and eyebrows get lighter, for instance, and those are both things that you could address with makeup, if you wanted."

The biggest surprise of all was the power of the brow. For women of all ethnicities, brow color faded with age—so darkening them may really make people look younger, the researchers say.

Though the study was only done in women, the findings likely apply to men, too. Other research suggests that the decline of facial contrast with age is not just true of women, but also true of men.

Stop Making These Eyebrow Mistakes That Are Making You Look Older

We fill you in on some common eyebrow mistakes that you may be making when shaping and maintaining your brows. Plus, how to fix them so you never end up with brow-don'ts ever again

A well-groomed set of eyebrows can help to frame your face and bring out your best features, but when they are too sparse, too dark, or over-dramatic, they can detract more than complement. Like the rest of your body, your brows can unfortunately start to show your age. They can start to get thin, become coarse and unkempt or turn grey.

If you've never paid too much to your arches besides the occasional plucking and tweezing, now is a good time to start. Just like a new hairstyle, a set of well-defined brows can trim away (literally) the years and help you to look younger than you actually are.

Arch nemesis #1: Thinning brows

Thick, bushy brows are a sign of youth. As you age, the natural ageing process thins the hair on the body whether it is the hair on your head or on your face. Even though we can't stop time, there's still hope. There are many restorative products that can help to bulk up the brows and repair overly-tweezed arches like brow enhancing serums and conditioning gels.

Alternatively, you can also create the illusion of fuller-looking brows with a brow powder or brow mascara to fill patchy spots. For a permanent solution, there is also microblading or brow embriodery that is similar to tattooing, but uses a thin microblade pen to create a natural-looking brow that can last up to three years.

Arch nemesis #2: Brows that are too dark

When it comes to your eyebrows, sometimes less is more. Overdrawn brows that are too thick and dark can look too harsh and heavy-handed.

As a rule of thumb, always stick to short, feathery strokes so you can build up the colour in a gradual way. It is also a good practice to take a step back away from the mirror to make sure you haven't gone overboard.

Arch nemesis #3: Over tweezed/plucked edges

If you cut into the natural arch of your brows, you might wind up with a choppy, uneven, unbalanced look. Rather than trimming the edges, brush the hair upwards and use a small pair of scissors made just for trimming the eyebrows to trim the brow hairs that stick out.

To help you gauge how long your brow hairs should be, hold a makeup brush handle or a comb with a pointed handle up to the highest point of your nostril, then follow it up to your brow. This is where your brow should begin. To find your arch, hold the handle and point it to the side of your nose, angling it so it cuts through your pupil. Next, move the handle to the side of your nose and angle it to reach the outer corner of your eye to determine where your brow should end.

Arch nemesis #4: Droopy brows

A brow that points down in the direction of your earlobes can cause the face to look saggy and make you look unhappy. To fix this problem, define the ends of your eyebrows first with short, feathery hair-like strokes that start from the base of your brow. Then, fill in the rest of the shape with an eyebrow powder (for a more natural finish) – working backwards towards the bridge of the nose.

Arch nemesis #5: Your brows do not complement your hair colour

To create a cohesive look, your brows and hair must work together. If you have jet-black hair, you wouldn't want to have bleached blonde brows for instance.

That said, you don't want your brows and hair to match exactly either because it will look unnatural. Consider buying a brow product that is a shade or two lighter than your hair colour. That way it looks more natural and less harsh.

Chapter 5

49 Classic Haircuts That Make You Look Like Like a Young Star

If you haven't heard, the term "anti-aging" is no longer part of our vocabulary here . It's dead and gone — yes, like that Justin Timberlake song — and with it, the archaic misconception that women can only wear a certain haircut at a "suitable" age. So go ahead and toss what society has told you out the window, because despite what you, me, and the generations before us were conditioned to believe, one's ability to own a specific cut or style has absolutely zilch to do with age. There are several other factors involved like hair type or texture and what face shape you have, but whether you're in your 20s or your 70s — consider that information N/A. Don't believe us? These celebrity cuts (plus the stylists who know their sh*t) are sure to prove otherwise.

Bangs: Long and Soft
FYI, all three hairstylists I spoke to waxed poetic about bangs, assuring me that they're one of the most complementary cuts out there for all ages. Who woulda' thought that everyone's hair nightmare from childhood could actually be so versatile? Well, certainly not me, but hey, you don't argue with the experts.

"Bangs are great for any age and for all hair types," says celebrity hairstylist and co-founder of MANE Society, Tippi Shorter. (Here, Dakota Johnson wears her fringe wispy and on the longer side.)

Hairstylist and salon owner, Nunzio Saviano, agrees, adding that he finds bangs to be the most flattering thing on every face shape as long as they're done correctly. "They're almost like an accessory for the face," he says. "They're ageless — any woman can have bangs."

Bangs: Sideswept
"Bangs and botox," celebrity hairstylist Matt Fugate says, clarifying that many of his mature clients have been coming in for fringe as a means of camouflaging any forehead wrinkles or fine lines they're not happy with. "They have a bad reputation from when you were little, but really any age can work them," he says. (Halle Berry is proof of this.)

Bangs: On a Curve
Goldie Hawn has worn her hair with bangs for pretty much the entirety of her career, and you know what? She's still slaying in them. Pro tip for mature women: Saviano advises against thick, blunt bangs or any sharp lines, and instead opting for a softer, piecey look like Hawn has going on here.

Bangs: Brow-Grazing

"Brow-skimming bangs work great on more mature faces," says Shorter, adding that longer styles look fresh on younger faces. Grace & Frankie star Jane Fonda has been nailing the same fringe for decades now.

Bangs: Blunt With Razored Ends
Fact: nearly-blunt bangs look so chic paired with a crisp, chin-grazing crop like Taraji P. Henson has here.

Bangs: Faux for Fun
If you've been on the fence about whether to get bangs or not, consider taking a cue from Gigi Hadid, who wore faux fringe to the MTV Movie Awards last year for a fun change of pace. (These are a fab and inexpensive option.)

Bob: Blunt and Textured
The bob is a classic for a reason: It works on anyone and everyone, simply because there are so many variations of it out there. "It also doesn't matter if you have fine or thick hair, as it's really all about the length and the hair all sitting in the same spot," says Fugate.

What's more, you can customize the bob cut for your face shape, says Saviano, whether it's chin-length, right below the chin, or grazing the ears.

The bob is basically as inclusive as a cut can get —
you just need to know what style flatters your features
the best. (Here, Lucy Hale wears hers textured and
right at the jawline.)

Bob: Choppy Layers
Rihanna wears her curly hair cropped with a sleek
middle part and face-framing angles.

Bob: Pin-Straight
If you have an oval face like Insecure star Yvonne Orji
then this super-sleek, pin-straight crop will look
lovely on you, as it helps sculpt the face and lift the
cheekbones — without having to whip out your
contour powder.

Bob: Long with Layers
Shorter says she loves to see more texture on mature
women, as it helps to ease any fine lines and it always
looks chic. Jada Pinkett-Smith has had both long and
short hair in recent years, but this style is easily one
of our favorites.

Bob: Angled
Fences star Viola Davis wears a sleek, angled bob
with a slight wave in the front, which Saviano says
adds a "softness" to the face in mature women.

Bob: Deep Part

Katie Holmes ear-grazing cut looks especially amazing paired with a bold side part, which elongates her face and neck. Tucking your hair behind your ear and letting the fuller side fall forward is another easy way to enhance your features.

Bob: Graduated
Our inimitable September 2017 cover star Helen Mirren — the Dame herself — wears her silver hair just below her ears with subtle sideswept bangs.

Bob: Curled With Bangs
Gayle King's piece-y fringe and soft curls draw attention to her eyes and frame her face in an ultra-flattering way that works at any age but is especially fabulous on mature women.

Curls: Rounded Cut
Curls, curls, curls! Shorter advises the best curly cut for all ages is around shoulder length with a touch of roundness to it. "It accentuates all face types and works well with waves, curls and coils," she explains. Black-ish's Yara Shahidi wears her natural curls on the regular, and never fails to elicit major hair envy when she does.

Curls: With Bangs
Zendaya wears her coils cut just below the shoulders with bangs, which, according to Shorter, has been extremely popular. "Curly bangs have been all the rage lately and work from loose curls to the tightest," she explains. Add extra shine to your strands with Vernon Francois's Pure~Fro Hold And Shine Serum.

Curls: Loose and Voluminous
Tori Kelly's blonde curls err more on the wavy side of the spectrum and have a fun, fresh-from-the-beach vibe that's ultra-flattering.

Curls: Blunt Lob
While Oprah often wears her hair straight these days, when she did fancy it curly, she was all about the strict-to-the-shoulders length and lots of volume.

Curls: Layered Lob
Actress and singer Margaret Avery wears her curly strands just above the shoulders and layered throughout.

The Lob: Piecey Ends
Otherwise known as the long bob, the "lob" is easily one of the most ubiquitously-worn cuts we've seen in recent years. And you know why? Because it's another cut that looks amazing no matter what your age. Take it from Shorter, too, who says, "[a] deep side-parted lob is a striking cut and works well to give anyone a polished look," adding that she loves to see a super sleek version on younger women and lots of texture on those who are more advanced in age. Selena Gomez chopped her long hair into a lob style last year, and we've been living for the look ever since.

The Lob: Blunt

Gabrielle Union regularly switches up her strands — because why not? — but sported this sleek, center-parted lob last year at a pre-party for the ESPYs. (Cue everyone swooning.)

The Lob: Allover Layers
Savaiano is another fan of the lob, touting it as a great cut for any face shape. What he does recommend, though, is that mature women keep things on the softer side — meaning no harsh middle parts or iron-straight strands. "Keep the softness and you can pull it off at any age," he says. Penelope Cruz has worn just about every hair length under the sun, but this shoulder-grazing cut is one we're most smitten with on the star.

The Lob: Face-Framing
Kelly Ripa fancies adding a bit of bend to her hair, which allows it to fall and frame her face in laid-back way.

The Lob: Soft and Wavy
Susan Sarandon nails the textured, "lived-in" lob look like no other. Try using Ouai's Wave Spray on damp hair and then diffusing to cop a similar look.

Midlength: Beachy Texture and Blunt Ends

Reese Witherspoon has been wearing her trademark blonde hair at this medium length — with few exceptions — for what feels like decades now. And you know why? Because it's timeless, that's why. Medium styles are extremely versatile since they can be worn in so many different ways, which is exactly why they complement so many people.

Midlength: Sideswept
Have a similar cut and hair type to Lea Michele? Try sweeping the majority of your hair to one side — eschewing a noticeable part — as this gives off an effortless vibe and adds volume.

Midlength: Straight and Sleek
Let it be known: Lily Collins is not one to shy from bold hair transformations (the actress has trialed pixies, lobs, and more). One universally chic cut we're partial to is this just-above-the-boob cut, which she wore pin-straight and shiny to an event last year.

Midlength: Tons of Texture
Julia Roberts is a glowing fountain of youth with skin-illuminating highlights and slightly gritty, just-back-from-the-beach texture. If your hair is at a similar length to hers, you can't go wrong.

Midlength: Long Layers and Side Bangs
Christie Brinkley is another celeb who's chosen to wear her sunny blonde hair in a similar style for years and years. For events, the model-slash-actress usually goes for loose waves with lots of volume — a look that literally works for everyone. No exceptions.

Waves: One Length
When it comes to long hair, there's a myriad of different lengths and styles you can shoot for — i.e. waist-length and straight, boob-grazing and tousled — though Shorter recommends waves with a center part for every age, as she says it offers a "super relaxed and youthful look."

Fugate also adds that as long as you have healthy hair, you can wear it long regardless of how old you are. Amanda Seyfried keeps her luscious strands shiny and sleek with a low-key loose curl to it.

Waves: Layered Throughout
Priyanka Chopra wears her medium length hair with long, face-framing layers, which highlights her prominent cheekbones. This style of cut works well on a gamut of face shapes, as the length is very forgiving. (Pro tip: When adding subtle waves to your hair, be sure to prime first with a protective spray or serum like Verb's Ghost Prep.

Waves: Long, Blunt Layers
Cindy Crawford's blunt ends help accentuate how thick and healthy her hair is, while shorter pieces in the front soften her angular features.

Waves: Long Choppy Layers
Pro tip: Incorporating layers into long hair (as seen on Regina Hall, here) is a foolproof way to add body and movement.

Waves: Long Pointy Layers
All hail Connie Britton, the long hair queen who's worn it boob-length or below her entire career — and somewhere along the way became famous for it.

Waves: Midlength With Layers
Michelle Pfeiffer's lightly-layered style falls just far enough below her shoulders to make it into the Long Hair Club.

Pixie: Shaggy
Fugate swears by the cropped cut, aka. the pixie, for women of all ages, as does Shorter. "It's a haircut that says strength, confidence, and fun" and "can work at any age and texture to exude the same feelings," she says.

Cara Delevingne shaved off her hair entirely for a film role, but it's since grown out a bit, giving her the chance to play with fun makeup looks and bold accessories.

Pixie: Long on Top
The singer-songwriter wore her pixie slightly spiked in the front, adding an edgy effect we're very much about. (Cop Halsey's rocker-chic look by using a tiny amount of Living Proof's Molding Clay to create a piece-y effect.)

Pixie: Sideswept
Swept oh-so-slightly to the side, Winnie Harlow wore her crop sleek and straight in Cannes this year.

Pixie: Classic
Jennifer Hudson's cute crop shows off her stunning features — the sideswept fringe creating the perfect frame for her gorgeous eyes.

Pixie: Purposefully Disheveled
Zöe Kravitz's spiked pixie, which is just a wee bit longer at the top and textured throughout, is the perfect way to style a short cut if you're into edgier looks like the Big Little Lies star tends to go for. It also brings attention to a bold brow and allows your beautiful features to take center stage.

Pixie: Soft and Layered
Truth be told, I can't imagine Ellen Degeneres without her signature pixie. It's clearly a cut that works for her and makes her feel like her best self, and that's all that matters, right?

Pixie: Elevated
"Whether you're 20 or 60 you can pull off a pixie cut," says hairstylist David Babaii of Kate Hudson's pushed-back pixie. While growing it out, styling it this way with a bit of texturizing paste takes seconds — and looks damn-good no matter your number.

The Shag: Long Layers
Oh, the shag. It's the ultimate "cool-girl cut," if you will, and that's likely because it typically looks great on a wide range of people, which Saviano says is because you can play around a lot with the layers. "You can have them longer and flatter or with more body and bangs if you want," he explains, adding that it doesn't matter if your hair is shoulder length or longer, and that it can be modified for any face shape.

Multi-hyphenate talent Alexa Chung is probably most well known for her signature shag, a French girl-inspired cut that makes her look the epitome of chic. Get a similar texture to Chung's by using Moroccan Oil's Dry Texture Spray.

The Shag: Short Layers
Taylor Swift's sideswept shag (paired with an icy dye job) elicited a completely edgy vibe out of the country-pop singer.

The Shag: Medium Layers
Diane Sawyer's midlength shag, which showcases both long and short layers is ideal for someone who wants to experiment with the rocker-chic style cut without going all in yet. Subtle side bangs also help soften features while making your eyes stand out.

Below the Shoulder: Straight and Layered

Not sure if you can pull off short hair just yet? (Spoiler alert: you can.) But if you're someone who's used to having long strands, we 100 percent understand the apprehension. That's why a midlength cut like Jennifer Aniston's, which falls slightly below the shoulders, is an awesome segway, as it's an ideal medium between lengths: not too short, nor too long. Oh yeah, and it works on just about everyone.

Below the Shoulder: Middle Part
Octavia Spencer's chic medium-length cut looks fabulous is accentuated with a crisp middle part — a style that looks especially great on rounder face shapes.

Below the Shoulder: Soft Waves
Laura Dern's classic collarbone-graving cut looks stunning styled in loose waves, with the hair facing away from her face to show off her features (those cheekbones, though).

Below the Shoulder: Deep Side Part
If you tend to go for a middle part, try experimenting with a deep side part like Rose Byrne did here, as it's a great way to switch up your look and see your features in a totally new way.

Chapter 6
Sleeping Yourself To Youthfulness Is Possible

We do so much to make our skin look great in the morning. Our bathroom counters are cluttered with everything from 10-step skin care to Fenty foundation, or the most recent Amazon haul from clean beauty brands.

But what if one of the biggest secrets to better skin was as simple as laying down and taking a nap? After all, our body never stops working — especially when we're asleep.

It turns out there's quite a bit of research and science behind the concept of beauty rest. Sleep is when some of the most important internal — and epidermal — recovery takes place!

While you shouldn't fully abandon your daytime skin care routine in favor of getting more Zzz's, there are some easy ways to amp your skin-sleep relationship for morning results.

How sleep affects your skin
You can almost immediately tell that getting a poor night of sleep doesn't do woke-up-like-this wonders for your face. Research even says that one night of poor sleep can cause:

- hanging eyelids
- swollen eyes

- darker undereye circles
- paler skin
- more wrinkles and fine lines
- more droopy corners of the mouth

A 2017 study found that two days of sleep restriction negatively affected participant's perceived attractiveness, health, sleepiness, and trustworthiness.

So, what seems like an overnight issue could transform into something more permanent.

First and foremost, you should understand that sleep is the time when your body repairs itself. This is true for your epidermis as much as it is for your brain or your muscles. During sleep, your skin's blood flow increases, and the organ rebuilds its collagen and repairs damage from UV exposure, reducing wrinkles and age spots.

Second, sleep is a time when your face inevitably comes into contact with the elements directly around it for a long time, especially if you're getting the recommended seven to nine hours each night.

Think about it: Your face against rough, drying cotton for one-third of its existence and being exposed to the sun for two unprotected hours could do a number on the appearance and health of your skin. Here's what you can do to help give your skin a rest.

1. Get a full night of sleep

The best place to start for your skin — and for your overall health — is to get the recommended amount of rest each night.

The results of poor sleep for your skin are numerous and significant, including:
skin that ages faster
skin that doesn't recover as well from environmental stressors like sun exposure
less satisfaction with your skin qualityTrusted Source
Sometimes you might have an off day but you should average seven to nine hours of sleep. If you're wondering how to reset your internal clock and catch up on rest, try sleeping in on the weekends by following our three-day fix guide.

You can also track your sleep with a wearable fitness tracker.

2. Wash your face before turning in
We've established how sleeping is a surefire way to help your skin repair itself: blood flow increases, collagen is rebuilt, and the muscles in your face relax after a long day.

But going to sleep with a dirty face can also harm the appearance of your skin.

Cleansing your face each night is arguably more important than in the morning — you don't need to use fancy products or scrub too hard. A gentle cleanser to remove dirt, makeup, and extra oil will do the trick.

You don't want to give the day's pore-clogging irritants the chance to sink in and do damage overnight. This can cause:

3. Use an overnight moisturizer and put a glass of water on your bedside table

Washing your face can dry it out and sleeping can also dehydrate skin, especially if you snooze in a low-humidity environment. While staying hydrated by drinking water can help to some extent,Trusted Source what your skin really needs at night is a topical moisturizer.

Again, you don't need the fanciest product on the market. You just need a thicker cream or oil that can help your skin as you sleep. Another option is to use your day moisturizer and layer petroleum jelly — using clean hands — on top to lock in the moisturize. For a more supercharged product, try an overnight sleeping mask.

4. Sleep on your back or use a special pillowcase

It makes sense that the position your face is in while you sleep (for one-third of your day!) matters to your skin.

Sleeping on a rough cotton surface can irritate your skin and compress your face for long hours at a time, resulting in wrinkles. While most wrinkles are caused by the expressions we make while we're awake, wrinkles on the face and chest can result from sleeping on our stomachs or sides.

An easy solution to this is sleeping on your back —
which also has a few other benefits — even if you have
to train yourself over time.

If you prefer to sleep on your side, get a skin-friendly
pillow. A satin or silk pillow minimizes skin irritation
and compression while copper-oxide pillowcases may
reduce crow's-feet and other fine lines.

5. Elevate your head
Elevating your head has been proven to help with
snoring, acid reflux, and nasal drip — all issues that
can disturb the quality of your sleep, and therefore
your skin. In addition, it can help reduce bags and
circles under your eyes by improving blood flow and
preventing blood from pooling.

Elevating your head while you sleep can be as simple
as adding an extra pillow, adding a wedge to your
mattress, or even propping the head of your bed by a
few inches.

6. Stay away from sun while you snooze
While we do most of our sleeping in the dark,
sleeping with your skin directly exposed to the sun in
the morning, or during naps, can have a damaging
effect on your skin's health and appearance — not to
mention that sleeping in a lighted room can disturb
sleep and sleep rhythms.

Getting blackout curtains or making sure that your
bed is out of the sun's direct line can help.

Embrace healthy sleep as a way to healthy skin
In 2019, the skin care industry will see an estimated
$130 billion dollars of global sales, in the form of
lotions, fillers, serums, and scrubs. But while we often
spend a lot of our time layering and lasering our skin,
paying attention to how we treat our skin during
sleeping hours shouldn't be overlooked.

It's not just for a glow or looking youthful, it's about
maintaining your health in body, mind, and skin for
years to come. A few wrinkles never hurt anyone — in
fact, they're usually a sign of happy years lived.

Chapter 7
Studies Prove Your Mindset Determines How You Age

The answer is "Yes" according to Professor Ellen Langer. During the last forty-five years, this Harvard social psychologist has studied the way our mindset affects both our health and how we age. At the core of her work is unifying the mind and the body rather than how the conventional medical and ps ychological world typically treats each as separate. Langer is convinced that a unity offers a far better understanding and hope for making positive change. Fortunately, her studies provide us with plenty of science to back up her assertions.

Ellen Langer teaches in the Psychology Department at Harvard University and is the first woman ever tenured there. She is the author of eleven published books and over 200 articles. Of all her work, the one titled ***Counterclockwise*** is the most influential and gets to the heart of her ideas about the mind/body connection.

Counterclockwise asks the question: Can you remember who you were and how you felt 20 years ago? And if yes, how might that influence your body and mind today? With those questions in mind, Langer and her students recruited two groups of older senior men in their late 70's to early 80s. Back in 1979, this was when 80 really meant 80—so these guys were truly old. None of these men lived alone making them dependent on either a family member or a nursing home facility. Many walked with a cane and all needed a support system for the majority of their needs.

Before the study began each man was carefully tested for what was considered to be biomarkers for age at that time—everything from memory and cognition, to flexibility, dexterity, grip strength, and of course their hearing and vision. Even their mental state was recorded.

Then Langer divided the 16 men into two groups and at separate times took them to a retreat center that had been carefully replicated to look exactly as it would have 20 years earlier—1959. Everything in the retreat center was meticulously designed to ensure that nothing in the house appeared older than 1959; the black and white television in the living area, the appliances in the kitchen, and the magazines on the coffee tables. All records in the record player came from 1959 or earlier, and all the TV programs and movies came from that earlier period. Mirrors were removed and only clothing of the era was allowed.

Each man was told in advance that they would be part of a study, but not that the study had anything to do with aging. It was explained to the first group that their mission was to reminisce about the past. In contrast, the primary group was instructed to *act as if* it was actually 1959 in every way. They were encouraged to psychologically attempt to *be the person* they were 20 years earlier. They were also coached to only talk about events and happenings that had occurred in the world or to them, prior to 1959.

Also important was the fact that both groups were treated as though they were 20 years younger. They were required to carry their own luggage, help with dinner and cleanup, and make up their own rooms. When the study concluded, they again tested all the men. Surprisingly, both groups of men (the control group and the primary group) registered noticeable improvement in some areas including hearing, vision, and memory. But, showing even more improvement were those in the primary group who also registered greater flexibility, faster gait, greater manual dexterity (where their fingers actually lengthened in spite of arthritis), and improved posture. Sixty-three percent of them scored higher on intelligence.

Also of note was the fact that although they had arrived extremely dependent upon either family or institutions to manage their needs, each man began functioning independently almost immediately upon arrival. Photos taken prior to the study and at the end, showing a visual difference. Independent observers rated the seniors, although still senior, as looking somewhat younger and more vibrant.

The results were so astounding that Langer hesitated to publish the outcome at the time, believing that she would not be taken seriously as a scientist if she did. However, ever since that time her research has continued to root out the numerous ways that our mindsets and thoughts influence our bodies.

The next study by Langer also confirmed the mind/body connection. Called the Chambermaid Study, this research shows that after hotel maids were educated to see how their daily actions could be perceived as healthy exercise—and with doing nothing different than just believing their work was indeed exercise—they lost weight, and their BMI and blood pressure improved. In other words, what people believe about their work and how they perceive exercise is connected to how their body responds.

Langer and her team then went on to study memory in a group of nursing home residents. First, everyone was given a memory test. Then ½ of the residents were asked to pay mindful attention to certain things in their home. To encourage that action they were offered incentives to recall certain things and events when asked. As Langer says, "Because they wanted the gifts, the information we asked them to track now mattered to them."

After a three-week period they found that *"when remembering mattered, memory improved."* But that wasn't the only benefit, those offered incentives and instructed to be more mindful also became more cognitively aware—they paid better attention to other people around them, their rooms, and the nurses—and even increased their longevity in the years following the study.

Langer also reminds us of other research that demonstrates how our use of certain words has **the power to "prime" us**. After reading words associated with being old and aged, looking at photos of older people or items associated with advanced age, or doing tasks that focus us on what our society thinks it means to be older, we can prime ourselves so that our bodies respond in a slower and more limited fashion. Everything from the speed of our movements, to our eyesight, memory, and cognitive awareness can be affected.

And what about time? In one study Langer showed that when people are fooled into believing they didn't get enough sleep, they did worse on memory tests. When they were fooled into believing they received more sleep than they actually did—they scored better. In addition, tests about blood glucose levels being affected by perceived time are also relevant. When subjects believed that time was faster than normal, their blood sugar spiked accordingly. When time was "slowed down," the blood sugar responded in kind.

These, and nearly all of the studies done by Langer and her students demonstrate that *if* you can effectively change the mindset or perception of a person, you can often influence some of the physical responses in the person as well. This applies to the health of the individual, as well as how they age. One of Langer's most well-known students, psychologist Beccy Levy, along with her colleagues claimed after a study, that "those who viewed aging more positively lived, on average, seven and a half years longer than those who were negative." Other research titled the Berlin Aging Study "found that dissatisfaction with aging was one of the principle factors in how long people live." Again, if we can adjust our mindset, we can influence our body in more ways than we normally realize.

Fortunately, as Langer asks, "if our beliefs have influence on our well-being, surely we can learn to influence our beliefs?" Of course, fundamental to that idea is that we must be willing to believe we have some control over our own health. How can we do that? Here are several things Langer believes is crucial:

- First, we must be mindful or aware of the world around us as much as possible. Langer recommends, "Pay attention to what is new."
- Notice differences and variables instead of loss or decrements. Just because something changes doesn't make it bad or wrong.

- Recognize that the world is designed by younger people with different capabilities—but rather than seeing lost capabilities as a physical problem—choose to see them as a design issue.
- Realize everything is contextual. Sometimes just changing the context opens up a world of possibility.
- Refuse to be merely a number or statistic. We are all unique and that includes what is happening to us on a physical level as well.
- Get second (or third) opinions on anything related to your health or important decisions—and then stay mindful and open about the answers.
- Refuse to be labeled—especially in a way that limits you or "primes" you to believe you can't do something.
- Counteract negative stereotypes for aging or health. Refuse to be boxed in. "Change the game."
- Stop associating pain or disability with age. Seek other explanations for what is happening and understand that issues can happen at most any age.
- Refused to be "over-helped." Helplessness and dependence interfere with both our mental state and often our physical state as well.

As Langer says, "our attitudes, ideas, and beliefs are at least as important to health as our diets and our doctors." Yet many of us continue to believe the stereotypes of aging or health conditions as one of decline, decay, and inevitable loss. Langer goes on to say, "Our mindless decisions—our deference to doctors' opinions, our willingness to accept diagnoses, even the way we talk about our illnesses—can have drastic effects on our physical well-being." For those of us who want to age in a positive and healthy way, it is SMART to remember how much our mindset plays in the process.

Chapter 8
Could a collagen drink REALLY make you look younger?

Collagen can now be stirred into your smoothie and coffee, but do these drinks really live up to their anti-ageing claims?

Our bodies are teeming with collagen that gives our bones strength and our skin elasticity while replacing dead skin cells to keep us looking fresh.

But as we age, we deplete our stores of the restorative protein, leaving us to seek out an alternative source.

Collagen injections have been a popular anti-aging procedure for decades, but now many companies are marketing the youth-restoring stuff in everything from powders to pill, serums and, now, smoothies.

As part Healthista's March anti-ageing special, Vanessa Chalmers asked the experts which ones might really work.

'Peanut butter is the glue that holds my body together,' is a favorite quote of ours at Healthista. But if we're talking scientifically, it's collagen, from the Greek word for 'glue' that actually makes up 30 percent of our bodies' protein structures.

It's found in our muscles, bones, joints, and tendons and because collagen contributes to healthy hair and strong nails, and makes up to 75 percent of our skin, it's long been the top dog in the pursuit of a more youthful look.

Even more so because the body's production of collagen starts decreasing from our mid-20s by about one percent per year.

But if we take the word of a growing number of supplements now available on the high street, the reduction of fine lines and wrinkles could be as simple as adding a scoop of flavored powder to your morning smoothie. Or even your coffee. Sound too good to be true? We thought so too, so we asked the experts.

Do collagen supplements work?

What isn't so new is collagen supplements in the form of capsules and one of London's leading dermatologists Dr Stefanie Williams is a fan.

At her clinic, Eudelo Dermatology & Skin Wellbeing, she often recommends a three month course to patients to support cosmetic treatments as well as taking it herself.

The body easily absorbs collagen supplements for quick rejuvenation, Dr Stefanie says

'A good collagen supplement can boost your own skin's collagen production', she says. 'Natural collagen is too huge a molecule to reach your skin intact after swallowing,' she explains.

'To be absorbed by the gut, it would have to be digested into smaller units. Collagen supplements contain small collagen fragments – peptides and amino acids – which are easily absorbed by the small intestine and distributed throughout the body via the bloodstream, where they remain for up to 14 days.'

 Studies do show an improvement to skin with the supplementation of collagen
It's not possible to reverse ageing (but the way science is going, who knows what miracles are on the horizon – that's one thing millennials can look forward to).

However, there are some studies that do show an improvement to skin with the supplementation of collagen.

In a double-blind placebo-controlled study on over 100 45- to 65-year-olds, researchers concluded there was a significant reduction of eye wrinkles in those who took a collagen supplement (called Verisol) daily for eight weeks, compared to those who took a placebo. And the effects continued four weeks later.

A similar study assessing 'skin elasticity, skin moisture, transepidermal water loss and skin roughness', found positive results, and a significant increase in 'collagen density, skin firmness and nasolabial fold depth' with the daily use of Pure Gold Collagen for 60 days.

How does collagen work in the body?

Looking at the collagen market is a minefield – or should I say a menu at a restaurant.

Marine collagen from fish claims to improve skin firmness in eight weeks, while Dermacoll,used by Dr Stefanie Williams, is made of cattle cartilage called bovine.

 Our skin responds by increasing its own collagen production
'Because there are suddenly unusually high amounts of collagen building blocks floating around, your skin is tricked into thinking there must be some breakdown – a major injury, perhaps,' says Dr Williams.

'Skin responds by increasing its own collagen production conveniently using the building blocks we've supplied,' she says.

Some experts aren't so convinced, such as London anti-ageing doctor, Dr. David Jack.

'I was once told by a sales rep for one of the brands that "the presence of the hydrolyzed collagen tripeptides in the blood tricks the body into thinking it is in an anabolic state, so it releases growth factors to boost collagen levels in the skin." I've tried to research this and can't find anything whatsoever to support it so I suspect it is just based on someone's wild guess,' Dr Jack says.

What's more, according some experts, if you're looking to improve the skin specifically, you won't get very far.

'If it is the case that these growth factors essentially stimulate a fibrotic reaction in the body, then why would it just be the skin? Would it not happen in muscle and other tissues too?', asks Dr Jack.

Well, incidentally, a lot of these products rave that their collagen product could also soothe achy joints, improve gut health, and enhance athletic performance, too.

Collagen, like all other protein, is broken down during digestion, according to Maira Silva, a graduate of Biochemistry from Kings College London.

'Collagen is not absorbed AS collagen, it is broken down into amino acids,' Silva says.

'But it does provide the constituents required to synthesize more collagen in the body. Not much is required because our own collagen is recycled but a small amount can help since some is always lost.'

Vida Glow is one of the latest trendy collagen powders that can be added to a coffee or shake +3

Vida Glow is one of the latest trendy collagen powders that can be added to a coffee or shake

What are collagen drinks?

If you're looking at labels of powdered collagen supplements, chances are you'll come across the word 'hydrolzsed'.

Broken down into smaller units, this type of collagen has become popular because it can be dissolved in both hot and cold liquids. Collagen coffee, collagen soup, or collagen smoothie, anyone?

Not only are these powders easier to use, they're also easier to digest, according to Silva. 'Collagen is a fibrous protein (rather than a globular one) and so digestion of its native form is difficult.

Hydrolyzed collagen on the other hand (also called gelatin), a by-product of collagen hydrolysis, is a much better form of collagen to be ingested because it's simply easier for the body to deal with.'

Do be aware though, we are talking about a a product that comes from cows, chickens, fish and other animals, so if you're vegan, it's not for you.

By blending your sachet contents with your favorite smoothie or water, the collagen remains stable in liquid
Dr Anita Sturnham, GP and dermatologist, agrees. 'In order for collagen in a supplement form to be active in the deeper dermal layers of the skin, the collagen has to survive the acid digestion in our stomach and then cross the intestinal barrier in our gastrointestinal tract, before it can reach the bloodstream.'

'Absorption studies confirm that within six hours of reaching the blood stream, 95 percent of the collagen has left the blood stream and can be found in the dermis, the layer of the skin where peptides are actually needed and where they can get to work to stimulate remodeling', she says.

Pills and tablets have an absorption rate of 20 to 30 percent in comparison, according to marine collagen-makers Vida Glow.

'Pills and tablets can take up to 40 minutes for the body to breakdown', they say.

Another difference between powder forms and capsule supplements is the 'bioavalibility', or 'freshness' to you and me.

'By blending your sachet contents with your favorite smoothie or water, the collagen remains stable in liquid for approximately 30 minutes, which is why pre-formulated liquid collagen is unlikely to be bioactive', says Dr Sturnham.

Vitness Berry Boost is a fruity-flavored powder you can add to a beverage   +3
Vitness Berry Boost is a fruity-flavored powder you can add to a beverage

Not just a fishy drink

What these new sachet drinks do offer are a much more glamorized version of the common supplement.

Up until now, a drink supplement has most often come as fishy-flavored granules thrown back in a shot.

But Vitness now do berry and matcha flavors which are rather pleasant, especially added to nut milk for a sweet milkshake, for about $55.

What's more, they are bursting with added vitamins and minerals such as astaxanthin, hyaluronic acid, acai berry, raspberry, maca, green coffee bean and vitamin C, which have been shown to promote healthy skin.

'There are a lot of nutrients inside of these drinks which are good building blocks for collagen', says Dr Maryam Zamani, a leading oculoplastic surgeon and aesthetic doctor.

'However most people who do take a collagen drink are also those who eat better, exercise and generally look after themselves, so its hard to say whether it's the drink that's doing that or a combination of many different lifestyle choices', she says.

Do collagen drinks work? The jury's out

If such supplements have such promising effects, dermatologists must be handing them out at their clinics, right?

They aren't locating the places you want them to work, like the neck, décolletage and hands
Dr Maryam Zamani, dermatologist
'I personally don't sell them in my clinic because they aren't enough good clinical trials that are definitive', says Dr Maryam Zamani.

'Even with the randomized testing it's difficult to be able to differentiate the changes irrespective of other outside factors such as stress, hormonal fluctuations, lifestyle choices such as smoking and drinking and the sun [and it's effect on the skin]. These aren't necessarily accounted for.'

Taking collagen orally has 'minimal if any effect and very temporary', says Dr Nick Lowe, a renowned London dermatologist.

'They aren't locating the places you want them to work, like the neck, décolletage and hands where women usually find they are ageing. That's my scientific rationale.'

Injections of your own platelets deliver the collagen directly to problem areas

What may work for targeting that, albeit at a far heftier price, is the use of a new cosmetic treatment called PRP – Platelet Rich Plasma, which Dr Nick Lowe has been using for a couple of years.

Original collagen injections made from protein and connective tissues from cows fell out of favor a few years back due to dodgy side effects and injections were taken off the market.

But PRP uses your own platelets in your blood plasma which contain high levels of growth factors and stimulate the skin cells to produce more collagen.

'That's also used to very successfully to increase hair growth in specific types of hair loss. If you correctly protect and maintain the skin, it can last a year or more.

'The good thing about it is the results look natural because you're not injecting anything like volume fillers. Production of your own collagen gradually ages so it's a very natural appearance.'

How can you boost collagen naturally?

The new drinkable skincare regimens can put you out $48 plus for a months supply, which isn't great news if you were already trying to cut back on your Starbucks morning coffee.

When it comes to the extortionate ones, skeptical Dr Jack says, 'Why spend $276 per month on what a simpler supplement [such as vitamin C] would do for less?'

Although the evidence that ingesting collagen has a positive effect on wrinkles and fine lines is 'not convincing', the evidence for foods and vitamins is, says anti-ageing nutritionist Rick Hay.

'Don't overlook the foods and vitamins which we have good research on. You may be better off taking ingredients that stimulate collagen production rather than straight collagen which may not directly impact on more formation', he says.

'Vitamin C has a really positive effect on collagen cross-linking', says Rick Hay, nutritionist. 'It becomes like a mesh which plumps and lifts the skin'. And a vitamin C-packed orange will only cost you the change in your pocket.

Hay delves more into the fruit bowl. 'Blueberries, raspberries, macqui berries and strawberries are filled with polyphenols and anthocyanidins, powerful antioxidants that can help fight free radical damage on the skin's surface', he says.

'Bitter orange and bilberry have, as the name suggests, a bitter quality. Skin health is linked to digestive health, digestive function and regularity, and the bitter qualities of foods such as these will help promote a healthy liver and gallbladder function, resulting in better fat metabolism and toxin elimination which means the skin won't need to work as hard.'

Dr Jack, adds, 'I really don't think it's of any benefit taking collagen in supplements over a good protein with a wide spectrum of amino acids'.

Or what easier way than making sure you're eating enough amino-acid high protein such as salmon, chicken, steak or plant-based sources (quinoa and soy products provide all of the essential amino acids).

Collagen and lifestyle

Diet and genetics are only part of the collagen puzzle. Other lifestyle factors including smoking and high amount of sun exposure contribute to depleting collagen levels, and 'air pollution can age you by ten years', according to Dr Daniel Glass, leading Harley Street Dermatologist from The Dermatology Clinic. 'If you are looking to slow down the ageing process, I would recommend that you protect your skin from excessive skin damage and pollution'.

If you are tempted by the collagen supplements, Dr. Glass recommends, 'the smaller collagen molecules are more effective, but this is difficult for a consumer to identify.

Therefore, I would suggest picking a supplement whose effectiveness is demonstrated in published clinical trials.'

Those that are successfully clinical trialed have a 'gold standard dose' anyway, according to Dr. Williams.

Therefore, 'my recommendation is 10g per day, that is 10,000mg. There are countless collagen drinks on the market, many of them too low in dose that contain, in my opinion, less effective types of collagen. it's also important to make sure the product is not stuffed with sugar or artificial sweeteners.'

Otherwise, we can only have faith in face yoga, slatherings of SPF, hydration and a healthy diet.

Chapter 9
Science-Backed Reasons Why Exercise Makes You Younger

Is exercise the fountain of youth? Research shows that exercise's powerful impact on our physical and mental health can in fact slow down the aging process. Here's how exercise makes you younger, one cell at a time.

Increases energy efficiency
Can running make you younger? One study suggests it might. A University of Colorado study found that older adults who regularly participate in highly aerobic activities (running in particular) have a lower metabolic cost of walking than sedentary adults.

What does that mean? As we age, if we are active, we maintain our 'horsepower,' or fuel economy. This suggests that those who exercise can maintain a better quality of life because of their ability to move around easily.

Makes our skin younger

By increasing blood flow, sending more oxygen and eliminating waste, exercise keeps skin cells healthy and vital.

In one study, researchers asked sedentary volunteers to work out for 30-45 minutes, twice a week, at 65 percent of their maximum heart rate. After just 12 weeks, tests on their non-exposed skin, which had shown normal signs of aging before the study started, resembled those of a 30- to 40- year old.

How? By increasing blood flow, exercise helps nourish skin cells and keep them healthy and vital by sending more oxygen to them and carrying away waste. Additionally, when we sweat, our pores open and release the build up inside of them. Sweat purges the body of toxins that would otherwise clog up pores and cause blemishes.

Improves posture
Due to muscle loss and bone density changes as you age, your ability to keep a healthy posture starts to decline.

By strength training either by using resistance bands, weights or aerobic exercise, such as swimming, you can rebuild muscle and prevent bone loss.

Taking care of your core and your spine has the added benefit of keeping your body and joints strong, and your taller posture will shave years off of your appearance.

Improves flexibility

Although any type of exercise, including both aerobic and anaerobic exercise, can improve our flexibility, yoga and pilates in particular are highly effective at increasing flexibility as we age.

Don't be one of the many people who avoid yoga because they are inflexible. That is the perfect reason to try it! By increasing your flexibility, you can reduce the risk of injuries – such as hip injuries – as you age, increasing your chances of living a longer, healthier life.

Boosts mental capacity

Studies show that regular exercise boosts the size of our hippocampus, the area of the brain that's responsible for learning and memory.

Numerous studies show that regular aerobic exercise boosts the size of the hippocampus, the area of the brain that's responsible for memory and learning. Those who exercise may literally have the brain of a younger person, making it easy for them to continue learning and maintain sharp mental health for years longer than those who are sedentary and do not get regular, aerobic movement on a daily basis.

Additionally, all types of exercise promote the health and survival of brain cells as well as the growth of new blood vessels in the brain.

Keeps our metabolism high

As we age, our metabolism naturally slows down. As the pounds creep up, we put our bodies at higher risk for diabetes, heart disease and other serious health issues.

The more muscle mass we have, the quicker our bodies burn calories. So by exercising and keeping your body strong, we are better able to maintain a healthy weight and reduce our chances of sickness and disease.

Slows cell aging

Exercise doesn't just make you look younger. Exercise makes you younger. How? By turning off the aging process in your chromosomes. In order to stay young, you have to keep your cells young. Researchers have found that exercise can keep DNA healthy and young.

Telomeres, the caps at the end of our chromosomes that are responsible for aging, get shorter as we get older. Recent studies have shown a link between regular exercise and lengthening of the telemores, which suggests that exercise may literally be able to slow our clocks down and help us live longer.

Relieves stress

If exercise is the fountain of youth, stress is its antidote. Experts say that one of the keys to living a long, happy, healthy life is the ability to reduce stress and anxiety and move forward after stressful life events.

A trait geriatricians call "adaptive competence" describes our ability to bounce back after something stressful happens. Without it, research shows that a high stress level can have an enormous impact on our longevity, shedding up to 33 years off of our lives. Yikes!

Exercise reduces stress, making it more likely that we will live longer – and happier.

Lowers cancer risk
Some studies suggest that regular, moderate exercise may reduce the risk of some cancers. One study showed that regular physical activity can reduce the risk for colon cancer in men by about 24 percent. Other studies show that regular exercise may reduce the risk of lung cancer by up to 20 percent. Plus, once diagnosed, exercise may help keep cancer from spreading.

Best Anti-Aging Exercise
Turning back the clock isn't just about wrinkles and fine lines. It's about staying fit — not just looking it — and keeping your body moving.

"A sedentary person hits peak muscle mass at age 20 and it starts declining from there," National Academy of Sports Medicine master trainer Rich Fahmy told The Huffington Post. "If you are active, however, populations as old as 60 have been able to maintain their muscle mass. Activity makes a difference."

We spoke with Fahmy, who has 16 years of experience as a personal trainer, and Barbara Bushman, a professor of kinesiology at Missouri State University, about the best ways to keep your body "young" as you age.

1. Focus on the major muscle groups.

"Larger muscle groups ... you see a decline in these faster," Fahmy says.

These are the muscles you use for basic motions like pulling, pushing and squatting.

Bushman says resistance training is important throughout your life, whether you're young or older. "Older adults who engage tend to have higher muscle mass and are stronger and leaner than their sedentary peers," Bushman said. "It's like win-win-win when we start looking at the exercise side of the picture."

A squatting exercise is what Fahmy recommends, bearing in mind your own personal fitness level. Any fitness regimen should begin with a baseline health and fitness assessment by a professional to prevent injuries. Also, you can use this online tool to help find an exercise program that's safe for you.

The good thing about squatting is that there are several variations depending on your fitness level. If falls are a concern, you can use a banister or rail at home to help stabilize yourself, he says. If you're more advanced, you can even add some weight, once you're comfortable. Aim to do some kind of resistance training, even just body-weight exercises, on two non-consecutive days per week, with sets of 10-15 reps.

2. Keep it simple.

Brisk walking or swimming a few times a week can reap major benefits. The great thing about walking is that it can be done nearly anywhere and is a good pick for anyone worried about their fitness.

Swimming also provides a good full-body workout, as it offers some resistance, with low impact.

Having an enjoyable workout that you'll stick to and make part of your weekly routine is more important than doing overly-sophisticated exercises, Fahmy says.

3. Focus on balance.

Falls are a major concern for adults over 65, and can lead to broken bones or head trauma. "One day you might notice you're more unsteady or you don't catch yourself as well as you used to," which is a scary prospect, Fahmy says.

There are a couple of exercises which can help you work on your balance. As with any exercise, it's important you're aware of your surroundings (things like equipment you could trip over at the gym) and your fitness level before you begin. Make sure you hold on to something to support yourself if you're concerned about falls.

One suggestion is the single-leg balance and reach. You can read more about it here. But for a simpler modification, Fahmy says it's as easy as learning to shift your weight from one leg to the other and then slowly being able to balance on only one leg.

4. Work on your back and core.

While we might be more concerned about a flabby midsection or with our aesthetic appeal, Fahmy says it's our spinal stabilizers that often decline without us noticing. "It's part of your inner musculature that keeps your spine in line ... the little balancing muscles designed to keep you upright, that start going if you're sedentary."

An exercise he recommends to strengthen them is the quadruped opposite arm-leg raise. You basically start at "tabletop position" on all fours, with a straight back. You then slowly extend one arm and the opposite leg, at the same time, to help focus on your coordination and strength.

5. Stretch it out.

Flexibility should be at the core of any exercise program, the experts say. "Because of the habits we've built over the years, the body gets good at what you repeatedly do," Fahmy says. This is important to focus on, especially if you have, and have had, a sedentary lifestyle or sit at a desk all day.

Bushman says it's important to include stretching at least two days a week and make sure you hold each stretch from 30 to 60 seconds.

The static pectoral stretch for your upper body and chest is recommended.

Chapter 10
If You Want to Shave Off 10 Years From Your Age, Just Drink Water

When we saw the ABC News Article on how a woman took TEN YEARS OFF HER FACE by drinking only water for 28 straight days, we tossed our coffee, poured ourselves a tall glass (or was it a huge jug?) of water, and we started our own investigation.

As it turns out, one of the best-kept secrets for healthier, younger-looking skin is as easy to find as turning the tap. Water makes up a large percentage of your body weight, and when you don't drink enough of it, it shows in your skin. Without enough water, your skin looks dull and rough. You'll add years to your apparent age and defeat the purpose of spa treatments or facials.

Consider these four facts about water:

FACT 1: YOU ARE MOSTLY WATER
You probably contain more water than anything else. On average, the human body is 60 percent water, but that proportion varies by age and body type. Some body tissues also need more water than others, and skin is especially susceptible to showing the effects of too little water. When you're dehydrated, your body finds the water it needs from within itself, robbing your skin of the moisture that keeps it looking younger to supply vital organs with essential water. The solution: Drink enough water to keep every part of you feeling hydrated and healthy.

FACT 2: WATER IS GOOD FOR YOUR TEETH

Drinking water leads to healthier teeth. When you drink more water, you drink less of the things that can spoil your smile, such as coffee and colas. By preventing mouth dryness, you also help keep teeth and gums healthy. Maintaining a beautiful smile is important to your health and your appearance, so opt for water over other choices and do your mouth a favor.

FACT 3: WATER IS THE ULTIMATE MOISTURIZER

Water helps moisturizer do its job. Dehydrated skin looks visibly parched. Skin loses its glow, reveals every fine line and turns rough when you don't drink enough water. Moisturizers can counteract some of that, but they can only do so much; even the best moisturizer only penetrates the uppermost layers of skin and can't reach deep into the dermis to do its work. To keep skin cells plump and youthful from the inside out, you need to replenish their water supply both internally and externally. Give your moisturizer a break and drink enough water to hydrate your skin's deepest layers.

FACT 4: WATER SUPPORTS COLLAGEN IN YOUR SKIN

Collagen contains water too. The connective tissues that give youthful skin its bounce and elasticity need water to function properly and repair themselves. Someone who's seriously dehydrated loses skin elasticity. When you choose water instead of other options with your meals and as a between-meal refresher, you help your body keep the collagen in your skin working as intended.

The Fountain of Youth may be mythical, but your water fountain could be a good substitute. Take your own 28 day challenge and see how your skin responds; you might just be amazed at how much of a difference water makes for the way you look and feel.

A recent article in the online publication Daily Mail showed how Sarah, age 42, took 10 years off of her face in just one month. How did she do it? She drank three liters of water a day for a month. That's it, no expensive beauty creams or spa treatments, she simply increased her intake of water. Sarah, 42, started drinking 3 liters of water per day hoping to solve her headaches and poor digestion. She never expected that the simple act of drinking water would take 10 years off her face.

Chronic dehydration robs the skin on your face of the water. When that happens, your facial skin loses volume and elasticity. Dehydrated skin damages more easily, so wrinkles and lines appear.

Your body also expels toxins through your skin. In a properly hydrated body, skin-bound toxins are flushed out when you sweat. If your skin is dehydrated, it won't sweat enough to release toxins, so they build up in your skin, which makes your skin appear aged and unhealthy.

Dehydration Facts
- One in 5 women drinks less than the recommended daily intake of water
- Every system and function in our body depends on water
- Water flushes toxins from vital organs
- Neurologist: Headaches and poor digestion caused by dehydration

Sarah found out that she wasn't drinking enough water after speaking with a neurologist. She suffered regular headaches and poor digestive health. She also talked to her doctor about what the neurologist told her. Her doctor told her she the same thing her neurologist told her: She should be drinking up to three liters of liquid a day.

Drinking water made a huge improvement in Sarah's skin tone and complexion. As Sarah says: "this demonstrates perfectly – and rather frighteningly – what a lack of hydration does to a face."

Five ways that your face shows that you may be dehydrated:

- Dark shadows under and around the eyes
- Wrinkles

- Shriveled Lips
- Reddish blotches
- Skin lacks any lustre. It looks dead.

30 Days to better health: What happened when Sarah decided to drink water

Week One: When Sarah started drinking water she weighed 119 lbs and had a 28-inch waist. Her urine color started out as dark yellow – another sign that she was dehydrated. She was worried at first that 3 liters of water was "an awful lot". Her doctor told her: "I suggest you have a big jug of water in the morning, then another in the afternoon and another in the evening." He also told her: "Your kidneys, which filter waste products from the blood before turning it to urine, will quickly feel the benefit, as they will be getting a good flush through."

Before Sarah started drinking water, she had to 'go' about 3 times per day. By the end of her first day of drinking 3 liters of water, she had to "go" a total of 6 times, and her urine turned clear. Sarah also noticed that her bowel movements, which she described as "sluggish" had become "much more lively"

Fact: Why Sarah's bowel movements improved: Drinking water encourages peristalsis – the movement of food and waste through the intestines. Also, proper hydration ensures that there is enough water to lubricate the colon, and a well-lubricated colon is a slippery place!

Joint Mobility: Sarah also noticed that her joints became more flexible when she started drinking enough water. She confirmed that it was because she was drinking more water when she asked Gemma Critchley of the British Dietetic Association, who confirmed that adequate intake of water provides lubrication to the joints

Does coffee, tea, or alcohol dehydrate you?
According to the British Nutritional Foundation: Moderate amounts of caffeine do not cause dehydration, so they do count towards your fluid intake. But alcoholic drinks do not hydrate you. Alcohol is a diuretic a substance that causes you to urinate. For every one alcoholic drink, your body can eliminate up to four times as much water.

Prevent hangovers by drinking water
Hangover headaches result from dehydration: the body's organs try to make up for a lack of water by stealing it from the brain, as a result of which it actually shrinks. Headaches result from the pulling on the membranes that connect your brain to your skull. Ouch. Luckily, I escape all this and wake up hangover-free.

For years I've been doing ten minutes of yoga every morning straight after I get up, but I've been feeling stiffer over the past six months. Yet since I started drinking more water my flexibility has improved. Gemma Critchley, from the British Dietetic Association, confirms that water helps lubricate the joints.

Week Two Sarah lost one pound, her waist size remained 28 inches. Sarah notes that her complexion improved, and her skin tone was evener. The wrinkles under her eyes became less prominent, and the shadows under her eyes began to diminish.

To lose weight: Don't substitute drinking juice for drinking water!
Sarah asked Gemma Critchley if the juice was an acceptable substitute for water. Critchley told her:"'If you drink a large glass of juice, you could be consuming more energy than you need," which Critchley warned would lead to weight gain.

Sarah's chronic headaches: Gone. Cellulite on thighs: Gone.
During week two, Sarah noticed to her delight that she hasn't had a headache for over a week since she started drinking more water. She also noticed that her bowels continued to improve. Sarah was concerned that drinking a lot of water would make her stomach feel bloated, but the opposite thing happened. She noticed that it had become flatter than usual. Her husband was delighted to observe that the cellulite on her thighs had vanished.

Week Three: Sarah's weight remained the same (118 pounds) but she lost a half inch from her waist (27.5 inches). Sarah's face continued to look more youthful. The dark rings and wrinkles under her eyes had virtually disappeared, and her skin looks plumper and better nourished.

Sarah also noticed that she no longer rubbed her eyes when she woke up. She also discovered that when she put on makeup, her eyes seemed to be less wrinkled, and her skin more elastic.

Drinking water helps you stay awake
According to Dr. Emma Derbyshire, a senior lecturer in nutritional physiology with Manchester Metropolitan University and an adviser to the UK Natural Hydration Council: "Our brain is 73 percent water, so poor hydration can affect how it functions. Dehydration can reduce our ability to concentrate as well as our cognitive performance."

You can do this! Next time you're feeling drowsy during the day, drink a glass of water instead of an energy drink, soda, or cup of coffee

Can you mistake thirst for hunger?
Drinking water before a meal is recognized as an effective weight loss strategy. This may be because a dehydrated body can mistake thirst for hunger. Studies show 37 percent of people mistake thirst for hunger.

Are you thirsty rather than hungry? Next time you feel hungry, reach for a glass of water instead of a snack, it may just satisfy your hunger.

Week Four brought more good news for Sarah. She lost another pound and another half an inch from her waist. Her face has now fully recovered from the effects of chronic dehydration. Sarah had this to say of her new youthful look:

I genuinely can't believe the difference in my face. I look like a different woman. The dark shadows around my eyes have all but disappeared and the blotches have gone. My skin is almost as dewy as it was when I was a child. The transformation is nothing short of remarkable.

Alkaline water best for hydration
If you are ready to put the healing power of hydration to the test, consider these facts about alkaline water:

- Tastes better than plain water – slightly sweeter and more refreshing
- Hydrates 17% better than plain water
- Detoxifies the body of toxic metals – plain water doesn't
- Supports good bone health
- Neutralizes acidity

Why better tasting water matters: If your goal is to drink more water than you're drinking now, the easiest way to do that is to make your water taste better. Alkaline water tastes better than the water you're drinking right now. In fact, Life Ionizers guarantees your water will taste better than the water you're drinking right now!

The benefit of improved hydration: Skin that is well hydrated looks healthiest, because it is. Hydration helps maintain your skin's elasticity and enables skin cells to discharge wastes efficiently. Two important hydration benefits inside the body include improved digestive function and lower blood pressure. In fact, a glass of water an hour before bed can prevent a heart attack or stroke. Dehydration, when you sleep, makes your heart work harder because the blood becomes thicker. Drinking alkaline water ensures optimal hydration throughout your body.

Detoxification: Research shows that alkaline water drinkers excrete 10 different metals in their urine that people who drink regular water don't. In fact, a study done in Sweden showed that alkaline water drinkers had significantly lower levels of mercury in their hair samples than people who drank acidic water. This led researchers to conclude: "Alkaline water may provide protection against the toxic effects of mercury."

Bone Health: Some of the strongest evidence of alkaline water benefits is for bone health. Multiple studies have shown that drinking alkaline water reduces the levels of two hormonal markers for bone loss. Maintained over the long term, reducing bone loss can lead to healthier bones.

Neutralize acidity: There are benefits to alkaline water's ability to neutralize acidity both in and outside of the body. Alkaline water has been shown to neutralize hydrochloric acid (stomach acid) and pepsin (an enzyme that triggers digestion). Excessive levels of acidity and pepsin in the stomach lead to the pain and damage of gastrointestinal reflux. Japanese studies have shown that drinking alkaline water can lead to relief.

Chapter 11
11 Foods That May Be The Fountain Of Youth

1. Eating sweet potatoes has lots of benefits, including keeping your skin firm.

Every time you eat a sweet potato, you do your whole body a favor, including your face. "Sweet potato is an incredibly healthy food filled with carotene (key ingredient for glowing skin), vitamin C and B6, biotin and fiber and a strong antioxidant," says Stephanie Middleberg, a New York-based dietitian and founder of Middleberg Nutrition. It's also a good source of copper, which helps your body produce collagen, making the tubers a cheaper and less painful delivery system than facial injections.

But how you eat your sweet potatoes matters too: Sweet potato fries are not necessarily a healthier choice than regular ol' French fries. "It really depends how they are cooked," says Keri Gans, another New York-based registered dietician and the author of the The Small Change Diet. "If they are deep-fried, then don't be fooled, you're still eating a French fry." Try making some baked sweet potato fries instead, so you get the nutrition without the unhealthy fats.

2. The antioxidants in berries can repair your skin and may even keep your brain young.

Anthocyanin, the antioxidant that gives many berries their eye-catching colors, can also protect your skin from UV rays. "Strawberries are also a good source of silica, a vital trace mineral for healthy skin and connective tissue," says Middleberg.

On top of that, researchers at Tufts University and University of Maryland, Baltimore County found that after feeding rats a blueberry-and-strawberry diet over two months, and then using irradiation to accelerate aging, the rats' brains had lower levels of toxins in them. "We are hoping [these results] will translate to human studies as well," said Dr. Barbara Shukitt-Hale, the lead investigator of the study.

Try these red, white, and blue berry yogurt popsicles for something sweet and healthy (or any of these delicious summer berry desserts if you want to indulge a little more).

3. Beans help your skin by reducing inflammation and improve your heart health with protein and fiber.

"Beans, in general, are amazing," says Gans. They are full of protein, fiber, and B vitamins, all of which do your body good. The protein and fiber are good for your heart health, and the B vitamins, Gans says, reduce inflammation of the skin.

While different kinds of beans have different levels of specific nutrients, Gans says to go ahead and just eat the ones you like. "But," she says, "I'm not saying eat refried beans," which can be high in fat. Instead, eat them plain or with your huevos rancheros.

4. Oats are also filled with skin-improving B vitamins.

"I'm a huge advocate of oatmeal," says Gans. Oats are packed with a number of important nutrients, including B vitamins, fiber, and protein. Those B vitamins are great for your skin, hair, and nails, but they also go deeper to nourish your nervous system. The fiber helps keep your digestive system chugging along, and the protein keeps your energy levels up.

But, Gans adds, "I'm not suggesting instant oatmeal that's flavored with sugar." Try these superfood-packed overnight oats instead.

5. Milk can give you vitamin D if you aren't getting it from time in the sun.

Our bodies need vitamin D, which we usually get from the sun, so that we can absorb calcium and maintain good bone health. Plus, according to a 2007 British study, higher levels of vitamin D can also improve the health of the DNA in our cells, slowing the aging process and protecting against age-related diseases.

"You really only need 10 minutes a day of sun to meet your vitamin D needs," says Gans. But for some people, especially in the winter, even that 10 minutes may be hard to get. For those cases, Gans recommends milk, since nearly all is fortified with vitamin D. Some orange juices and non-dairy milks are as well. Just check the nutrition label.

6. Curry powder may prevent the growth of toxic brain cells and prevent Alzheimer's.

Curry has long been used in India to treat a variety of diseases, including breast cancer, liver disease, and hemorrhoids, but recent research shows it might also prevent Alzheimer's. In 2001, researchers at the University of California, Los Angeles found that curcumin, a compound in curry, lowered certain toxins in mice with brains modified to model the effects of Alzheimer's, concluding, "this Indian spice component shows promise for the prevention of Alzheimer's disease."

Try these easy roasted curried chickpeas, which come with lots of healthy protein and fiber too.

7. Sardines and anchovies are full of brain-healthy omega-3 fatty acids (and have lower mercury levels and price tags than most salmon).

Consuming more omega-3s in your diet can slow the natural deterioration of your brain as you age, according to a January 2014 study.

Look for European anchovies from the Adriatic or sardines from the Pacific to get the most omega-3s with the least mercury. (Bonus: The tiny bones in sardines are another good source of vitamin D.) Scared to try anchovies? Ease yourself in with a green goddess dip.

8. Cocoa may lower your risk of heart disease and prevent diabetes and cancer — but choosing your chocolate carefully is essential.

Not all chocolate is equally healthy, but the right kind can actually improve your health. "Pure cacao," Middleberg says, "is extremely healthy." In a study of the Kuna Indians off the coast of Panama, scientists attributed the population's low risk of high blood pressure and cardiovascular disease (despite their weight and salt consumption) to their high cocoa intake — 10 times more than what an American would typically consume.

However, those scientists also drew a distinction between Kuna cocoa — made from dried, ground beans with just a bit of sweetener — and commercial chocolate, which is usually highly processed and has lost much of its healthfulness. "You want to look for varieties [of powder, nibs, or chocolate] that are above 75% cacao," Middleberg says, "because the more cacao, the less sugar and fillers."

Eat cocoa nibs straight up if you like the bitterness, or try making these raw cocoa, nut, and berry truffles.

9. Healthy fats like olive oil and nuts can lower your risk of heart attack, stroke, and death from cardiovascular disease.

Not all fat is bad! Some kinds may even help you live longer. In a five-year study of almost 7,500 participants, Harvard researchers found that people who consumed either an extra liter of extra-virgin olive oil every week or an extra ounce of mixed nuts every day had significantly fewer strokes, heart attacks, and deaths than the participants on the low-fat diet. "The reduction was impressive," wrote Dr. Anthony Komaroff, editor in chief of Harvard Health Publications.

Nuts and olive oil are both easy to incorporate into your diet. Cook with olive oil and snack on raw nuts next time you find yourself reaching for potato chips. However, Middleberg says, "It is still easy to overdo it." She recommends limiting servings to 1 tablespoon of olive oil, 1 ounce of nuts, and 1–2 tablespoons of nut butter.

10. Spinach and other green leafy vegetables can boost your energy, make your bones stronger, and protect your eyes.

It seems like there's not much spinach can't do. Those little leaves are full of iron, vitamin K, and the nutrients lutein and zeaxanthin. The iron will boost your energy and build your muscles, while the vitamin K will help your bone density and prevent prevent fractures and osteoporosis. The lutein and zeaxanthin allow your retinas to absorb damaging light before it can hurt your vision, and lower the risk of chronic eye diseases like cataracts and age-related macular degeneration (AMD), according to the American Optometric Association.

Adding spinach to your diet is easy because you can eat it raw or cooked, sautéing it takes about a minute, and you can throw it into dishes where it's not even listed as an ingredient. To get the full benefits of fat-soluble vitamin K, be sure to eat it with some fat, like olive oil or cheese, so that your body can absorb it. Try putting it in a quiche like this one.

11. Moderate intake of red wine can improve cardiovascular health, may prevent cancer, and may even help treat diseases like Alzheimer's and Parkinson's.

Good news for red wine lovers: A glass a night can improve your cardiovascular health. According to studies from the University of California Davis, antioxidants in the red grapes' skin and seeds can reduce production of LDL ("bad") cholesterol, increase HDL ("good") cholesterol, and reduce blood clotting. Other studies have found that the antioxidant resveratrol in the grapes' skins can prevent tumor development, as well as help your body create nerve cells, which may help treat neurological diseases like Alzheimer's and Parkinson's.

Choose drier (less sweet) red wines for the biggest health benefits, say UC Davis researchers. Not a wine drinker? You don't need to pick up the habit. "If you're not drinking alcohol, I don't recommend you start just to get the health effects," says Middleberg. Eating grapes will give the same benefits, says Gans.

Chapter 12
How Stress Quickens the Aging Process

At any age, stress is a part of life. Young and old alike have to face difficult situations and overcome obstacles. While young adults struggle to establish a career, achieve financial security, or juggle work and family demands, older people may face failing health or dwindling finances -- or simply the challenges of retaining their independence. Unfortunately, the body's natural defenses against stress gradually break down with age. But you don't have to give in to stress just because you're no longer young.

Many seniors still manage to sail through their later years. "Successful agers" tend have a few things in common: They stay connected to friends and family, they exercise and keep active, and, above all, they find ways to both reduce and manage the stress in their lives.

The stress alarm

Stress comes in two basic flavors, physical and emotional -- and both can be especially taxing for older people. The impacts of physical stress are clear. As people reach old age, wounds heal more slowly and colds become harder to shake. A 75-year-old heart can be slow to respond to the demands of exercise. And when an 80-year-old walks into a chilly room, it will take an extra-long time for her body temperature to adjust.

Emotional stress is more subtle, but if it's chronic, the eventual consequences can be as harmful. At any age, stressed-out brains sound an alarm that releases potentially harmful hormones such as cortisol and adrenaline. Ideally, the brain turns down the alarm when stress hormones get too high.

Stress hormones provide energy and focus in the short term, but too much stress over too many years can throw a person's system off-balance. Overloads of stress hormones have been linked to many health problems, including heart disease, high blood pressure, and weakened immune function. For older people already at heightened risk for these illnesses, managing stress is particularly important.

Over time, the brain can slowly lose its skills at regulating hormone levels. As a result, older people who feel worried or anxious tend to produce larger amounts of stress hormones, and the alarm doesn't shut down as quickly. According to a study published in the journal Psychoneuroendocrinology, women are especially susceptible to an overload of stress hormones as they age. The study found that the impact of age on cortisol levels is nearly three times stronger for women than for men.

The flow of stress hormones can be especially hard on older brains in general. According to a report from the University of California at San Francisco, extra cortisol over the years can damage the hippocampus, a part of the brain that's crucial for storing and retrieving memories. Several studies have found that high cortisol goes hand in hand with poor memory, so we might be able to chalk up certain "senior moments" to stress.

Years of emotional distress may even increase the risk of Alzheimer's disease. A five-year study of nearly 800 priests and nuns published in the journal Neurology highlighted this potential hazard. The subjects who reported the most stress were twice as likely as the least-stressed subjects to develop the disease.

Speeding up the clock

Stress doesn't just make a person feel older. In a very real sense, it can speed up aging. A study published in the Proceedings of the National Academy of Sciences found that stress can add years to the age of individual immune system cells. The study focused on telomeres, caps on the end of chromosomes. Whenever a cell divides, the telomeres in that cell get a little shorter and a little more time runs off the clock. When the telomere becomes too short, time runs out: The cell can no longer divide or replenish itself. This is a key process of aging, and it's one of the reasons humans can't live forever.

Researchers checked both the telomeres and the stress levels of 58 healthy premenopausal women. The stunning result: On average, the immune system cells of highly stressed women had aged by an extra 10 years. The study didn't explain how stress adds years to cells making up the immune system. As the study authors write, "the exact mechanisms that connect the mind to the cell are unknown." Researchers do have a not-very-surprising theory, though: Stress hormones could be somehow shortening telomeres and cutting the life span of cells.

Stress management: The real fountain of youth?

The good news is that we can put what we know about stress and aging to work for us. Learn to manage and reduce your stress load and you have a better chance to live a long, healthy life.

Maintaining a positive outlook is one key -- a study by Yale University found that people who feel good about themselves as they get older live about seven and a half years longer than "glass half empty" types. Researchers say the people with more positive attitudes may also deal with stress better and have a stronger will to live.

Staying close to friends and family is an excellent way to cut down on stress. As reported by the American Psychological Association, social support can help prevent stress and stress-related diseases. The benefits of friends and family can be especially striking for seniors. An article published in the American Journal of Health Promotion notes that social support can slow down the flow of stress hormones in seniors and, not coincidentally, increase longevity. Other studies have found that social interactions can help older people stay mentally sharp and may reduce the risk of Alzheimer's.

Exercise, a proven stress-buster for people of all ages, may be especially valuable in later years. Regular walks, bike rides, or water aerobics can do more than keep a person strong and independent; exercise can actually help block the effects of aging on cortisol levels. A recent study published in Psychoneuroendocrinology found that physically fit women in their mid-60s had essentially the same response to stress as a group of unfit women in their late 20s. In contrast, women in their mid-60s who weren't physically fit released much larger amounts of cortisol in response to stress.

In the end, anything that reduces unnecessary stress will make the later years more enjoyable. Some people simply need to stop trying to do too many things at once. Others may want to try breathing exercises or other relaxation techniques. Still others may need to talk to a psychologist to find a new perspective on their lives.

Whatever the approach, fighting stress overload is worth the effort. The American Psychological Association reports that reducing stress in later years can help prevent disabilities and trips to the hospital. And if people end up feeling younger, healthier, and happier, that's OK, too.

Chapter 13
5 Vitamins Packed With Age-Fighting Power

If you're looking for the fountain of youth, stop walking down fancy department store beauty aisles and start taking a look at your diet. While topical creams may be effective in fighting wrinkles, sagging skin and other telltale signs of aging, dermatologists say what you put in your body is just as important as what you put on your skin.

I spoke with dermatologist Patricia Farris of Old Metairie Dermatology in Metairie, Louisiana to get the scoop on what changes you can make in your diet to slow the aging process.

"I'm sort of a whole food person myself," Farris, 61, said. "I always tell people, the best way to get any phytonutrient, vitamin or antioxidant, is in the foods it's been grown in." While taking supplements is a hot trend, Farris says there's no guarantee vitamins will have the same effect in supplement form as they do in their natural form.

"I you start beefing up your diet with heavy loads of fruits, veggie and antioxidants, you certainly could thwart some of the damages of aging," Farris said. "Just because you're over 50 doesn't mean you can't make a difference now. Clean up your diet and put in the good food no matter what age."

Here are five vitamins and their natural food sources that you might want to add to your diet to battle aging:

1. Vitamin C

Found in abundance in citrus fruits, vitamin C is essential for collagen production, skin repair and also for keeping your bones and teeth strong. Vitamin C not only protects against sun damage, it also can repair the existing damage which results in lines and wrinkles. "Vitamin C can change the way your skin ages," Farris says.

2. Vitamin A

sweet potatoes

Load up on foods like sweet potatoes, kale, mangoes and carrots to get the antioxidant benefits of vitamin A. Retinol, one of the only FDA-approved wrinkle treatments, is a form of this vitamin, which works inside cells, hitting receptors which helps them function as if they were younger. It also slows the breakdown of collagen, Farris says, which is important in giving skin its youthful appearance and elasticity. While it's important to get your vitamin A in your diet to keep a strong immune system and keep our skin and eyes healthy, it can also be applied topically for anti-aging benefits.

3. Vitamin D

When it comes to anti-aging, don't neglect your bone health. Bone loss can be one of the more powerful effects of aging to consider, especially for women, Farris says. Vitamin D works with calcium to keep your bones strong and dense. "One of the things that makes your face look saggy is loss of bone. We focused for years on collagen, and now we understand that you lose fat, collagen and bones in aging," she said. "The better your bone health, the better your bone structure is and the more youthful you will appear."

Foods like fatty fish, some dairy products and egg yolks are good sources also. According to the NIH, most Americans get the majority of their vitamin D from fortified foods, like most milk and some orange juices.

While vitamin D synthesis also occurs from our daily sun exposure, sunscreen with SPF 8 and higher can block that effect. The American Academy of Dermatology recommends getting vitamin D from your diet, as unprotected sun exposure can damage your skin and cause premature aging.

4. Vitamin K

Load up on leafy greens to get your vitamin K, especially if you complain about bruising. Farris says one of the most common complaints she gets from her post-50 patients is of bruising, commonly on the arms. Vitamin K helps keep collagen in your skin intact and keeps your skin thick, reducing the appearance of bruises, veins and, as some suggest, even dark circles.

5. Vitamin E

Vitamin E is great, especially when combined with vitamin C, in keeping your cells healthy. Vitamin E helps fight damaging free radicals on a cellular level, which protects your cells from vulnerability. Farris says it's one of the most potent antioxidants out there. It can help provide sun protection to the skin and may have anti-inflammatory benefits. It is fat-soluble and can be found in many oils, such as sunflower oil and soybean oil, as well as in nuts and seeds.

Chapter 14
5 Power Minerals to Look Young & Vibrant

While there are a ton of state-of-the-art, anti-aging skin care products and treatments on the market, looking younger begins from the inside out. You really are what you eat. Your food should not only satiate your every taste bud and be your fuel, but you should also think of your food as part of your skin care routine. anti-aging minerals into your breakfast, lunch and dinner will help yield more radiant, youthful skin.

But how do you do that? There are a ton of diets out there for this, that and the other, offering false promises, so it's difficult to know which ones to trust and which ones are bogus. So we called on expert nutritionist, Joy Bauer to give us some skin care tips. She is the nutrition/health expert for "The Today Show" and she advises everyone from the New York City Ballet ballerinas to famous actors and Olympic athletes. Here she shares what the top age-fighting minerals are, and explains which foods contain these minerals and how much of these foods you need to eat to get optimum benefits. You might be surprised to find out that some of your favorite foods are already packing a big anti-aging punch, but if not, her advice might inspire you to incorporate a few new foods into your next meal. See anti-aging minerals now.

- Calcium

The mineral calcium is well-known for its key role in bone health, teeth and bodily organs, including the skin, where it regulates skin's many functions. Most calcium in the skin is found in the outermost layer of skin (the epidermis) where it has been demonstrated to play an important role in barrier function repair and skin homeostasis 1 (self-replenishing process where the number of cell divisions within the skin compensates for the number of cells lost 2.) Within the epidermis, keratinocytes have a different need for calcium concentrations.

Despite the continuous renewal of the epidermis (almost every 60 days the epidermis renews itself completely, replacing more than 80 billion keratinocytes in the body of an average adult) our skin eventually succumbs to aging, as the turnover rate of the keratinocytes slows down dramatically. Aging is associated with thinning of the epidermis, elastosis, loss of melanocytes associated with an increased paleness of the skin and a decreased barrier function. As the differentiation of keratinocytes is strictly calcium dependent, calcium also plays an important role in the aging epidermis. Recently it has been shown that the epidermal calcium gradient in the skin that facilitates the proliferation of keratinocytes and enables their differentiation is lost in the process of skin aging.

- Chloride

SODIUM CHLORIDE SOUNDS SCARY, BUT IS IT?

Sodium Chloride is commonly referred to as "TABLE SALT"! Nothing scary about that right? Perhaps only in the way sodium chloride sounds that you would think it's a chemical made in a lab. In terms of safety (including the EU) this ingredient is considered safe for use in beauty products.

SODIUM CHLORIDE IS ESSENTIAL FOR OUR BODY

Sodium Chloride occurs naturally in seawater and is mined as mineral halite. Sodium Chloride is essential to our body and is found naturally in our body tissue and bodily fluids. Our body relies on the role it plays to transport nutrients and waste, as well as our electrolyte balance and nervous system.

FOR SKIN & SCALP

It has been used in the beauty industry for centuries and has a number of different purposes, some not always suitable for everyone. My favorite is the use of salt as an antiseptic as it can be a very healing and effective odor eliminator. Going for a swim in the ocean is often healing for most skin and scalp conditions as Sodium Chloride can be used in very "medicinal" ways.

IN FOOD

Sodium Chloride in food has been used for centuries as one of the most useful preservatives and antiseptics. We all know that it is used to prevent bacteria, particularly meat; ex. prosciutto is cured with salt in order to prevent bacteria.

REMEMBER, EVERYTHING IN MODERATION

Like all things in life, sodium chloride needs to be used in moderation both in beauty and in food. The more you know about its' uses, the more you can benefit from it. Sodium Chloride has been used for so long that we have very extensive research and data on how it affects your body inside and out. With its antiseptic and healing properties, this is why I like to use this ingredient in product development.

WHERE SODIUM CHLORIDE IS USED IN BEAUTY PRODUCTS

HAIRCARE

Used as a stabilizer, thickener, and sometimes pH adjuster.

It is often found in shampoos and conditioners. Check the label for the amount. If you see it in the top 5 ingredients (possibly as high as 20%), BEWARE. If it is at the bottom of the label, it is less of a concern (as it is at lower levels).

Salt at higher levels can be drying to your hair and scalp. Conflicting information exists; as I know many DIY recipes suggest that salt helps absorb excess oil, unclog follicles, manage sebum and product build. It also reduces inflammation associated with dandruff and psoriasis on the scalp. Higher salt levels can actually help scalp issues but only when used once per week, as it can have the opposite affect on your hair and scalp when over used.

SOAP

The addition of sodium chloride forms fatty-acid salts in a soap formulation. The key in soap formulation is to balance the sodium with oils in order to ensure that it is not too drying. Sodium chloride when used in a shower soap can have many medicinal and healing benefits as it can help treat some skin conditions and help exfoliate. The basic rule of thumb is salt should be one of the lowest on the ingredient list to avoid the drying effects on skin. Percentages matter.

NATURAL DEODORANT

Sodium chloride has been used as a simple solution to combat underarm odor. Some natural deodorant come in the form of a crystal rock, made of sodium chloride. Crystal rock deodorants work for some people and can but used to deodorize.

I did not use it in our natural deodorant because I found more powerful natural ingredients for a stick deodorant. However, I love the natural ability of sodium chloride to eliminate bacteria (which causes odor) and this is why I chose to put it in our underarm bar. We balanced it with apple cider vinegar, natural oils and essential oils in order to eliminate odor-causing bacteria under your arm. (I would not suggest it for any other part of your body, other than underarms and feet as it is drying).

Showering with the underarm bar leaves a sodium barrier on the skin which ensures no odor break-through, even for those who have a problem with excessive sweating and body odor.

ORAL CARE
Sodium Chloride polishes teeth, reduces oral odor, and cleanses and deodorizes teeth and mouth. Sodium Chloride also imparts a flavor or a taste to a product.

TIP: Try a good old warm water salt rinse in your mouth when you want to kill bacteria, instead of a commercial mouth wash.

- Copper

Copper is a trendy skin-care ingredient, but it's not actually anything new. Ancient Egyptians (including Cleopatra) used the metal to sterilize wounds and drinking water, and the Aztecs gargled with copper to treat sore throats. Fast forward thousands of years and the ingredient is making a major resurgence, with creams, serums, and even fabrics popping up with promising anti-aging results.

Today's creams feature a natural form of copper called copper tripeptide-1, says Stephen Alain Ko, a Toronto-based cosmetic chemist who has studied copper. Also called copper peptide GHK-Cu, the copper complex was first uncovered in human plasma (but it's also found in urine and saliva), and is a type of peptide that seeps into skin easily. Many of the newer products use these types of naturally occurring peptides or copper complexes, he adds.

Previous forms of copper were often less concentrated or irritating or unstable. Copper peptides, however, rarely irritate the skin, which makes them a popular ingredient when combined with other so-called cosmeceuticals (cosmetic ingredients said to have medical properties), says Murad Alam, M.D., professor of dermatology at Northwestern University's Feinberg School of Medicine and a dermatologist at Northwestern Memorial Hospital. "The argument for copper peptides is that they are small molecules important for various body functions, and if they are applied to the skin as topicals, they can enter the skin and improve its functioning," he explains. This translates to anti-aging perks. "Copper peptides may reduce inflammation and speed up wound healing, which may help the skin look and feel younger and fresher."

Before you stock up, it's worth noting that there's no conclusive evidence of its efficacy yet. Studies are often commissioned by the manufacturers or done on a small scale, without peer review. But "there have been a few human studies on copper tripeptide-1 on skin aging, and most of them have found positive effects," Dr. Alam says. Specifically, a handful of studies showed that copper may make skin more dense and firm, he says.

Dr. Alam recommends trying out a copper peptide for one to three months without changing other parts of your beauty routine. Keeping the other products to a minimum can better help you track skin results to gauge whether "you like what you see," he says.

- Iodine

Iodine is a component that you must've heard or studied about back in the day. But, unlike vitamins and minerals, iodine is always ignored and forgotten about. We never consciously think that "Oh, I need to get my daily dose of iodine", like we do for vitamins. But guess what? You should! Iodine is an essential compound that controls the functioning of the thyroid gland. It is responsible for our metabolic processes and the smooth functioning of our immune system. Even when it comes to skin, iodine is essential.

Iodine encourages detoxification in our bodies, thus preventing chemicals like chlorine etc. from harming the thyroid functionality. Iodine deficiency can have adverse effects on our bodies and skin.

Apart from regulating skin's moisture levels, iodine also aids healing of scars, cuts etc. Basically, it helps in skin repair. It helps in regeneration of the lower layers of your skin by triggering cellular function. This results into complete rejuvenation of your skin and hair and nails! Iodine also helps regulate the hormones that are responsible for acne breakouts.

Now that you know the numerous benefits of iodine for healthy skin, here are a few ways you can get your daily intake...
Consuming foods rich in iodine will prevent iodine deficiency, which is extremely common and often ignored. We recommend eating foods that have iodine as a naturally occurring substance as opposed to taking supplements; you can do so only when an expert asks you to. Consult with your physician / skin expert as to how much of iodine is essential for your daily intake and incorporate foods rich in iodine in your diet.

- Iron

Iron supplements are most commonly recommended for symptoms of iron deficiency, such as fatigue, dizziness, rapid heartbeat, and shortness of breath. However, boosting iron in the diet also has some beneficial "side effects" that helps one to look and feel better, both outside and in.

Hair, skin, and nails require iron to retain their natural luster, shine, and moisture. A simple blood test from a doctor will indicate whether an iron deficiency is to blame for hair, skin, and nails being brittle and dry. Here is a closer look at how iron can help boost the health, feel, and appearance of your hair, skin, and nails.

Iron and Skin Health
Pale, dull, and lifeless skin is a common symptom of iron deficiency, which can cause self-consciousness and insecurities about appearance and looks. Iron-deficient skin often appears to be pale, yellow, or sallow in color. Another skin-related condition associated with an iron deficiency is unexplained bruising.

When iron is consumed through foods or supplements, the body's cells absorb it the gastrointestinal tract. Iron is released into the blood stream, then attaches to a protein and delivers iron to the liver. Iron is needed to make red blood cells and bone marrow, and this whole process going on inside affects how the skin looks on the outside.

Iron for Hair Growth: How Iron Benefits Hair Health
If there is not enough iron in the diet, hair may become dry, lack shine, and even begin to fall out. A protein called ferritin is known to cause these problems because it is essential for the process of storing and releasing iron to all the parts of the body over time.

When the body retains an adequate amount of iron and other nutrients, hair should have its natural shine. Another important nutrient for hair health is zinc. Some food sources rich in both iron and zinc include lean red meat, lentils, and soybeans. An easy way to boost the health of hair is taking an iron supplement.

Iron and Nail Health
Nails are comprised of keratin, which are hard layers of protein that form to keep soft tissues safe and protected. When there is not enough hemoglobin in the body, nails don't get enough oxygen to stay healthy.

In addition to maintaining body functions like producing hemoglobin to carry oxygen to organs, iron is a necessary nutrient for maintaining healthy nails as well. When the body is not absorbing enough iron, nails will appear brittle and dry. Brittle, dry nails are prone to breakage, which can be painful, unsightly, and hinder the ability to work.

More Tips for Healthy Skin, Hair, and Nails
Once the body is getting the iron it needs to keep all parts of it healthy and functioning properly, there are other things that can be done to stay looking and feeling vibrant. Here are some more tips for maintaining healthy skin, hair, and nails, regardless of what season of the year it is.

- Consume foods rich in iron, zinc, and protein

- Avoid high-glycemic foods, like white breads and pastas
- Limit sugar intake
- Drink lots of water throughout the day
- Exercise every day to promote blood flow and overall health
- Don't smoke and avoid environmental toxins as much as possible

If you think your hair, skin and nails can benefit from an increase in iron, talk to your doctor about taking an iron supplement, such as Fergon.

- Magnesium

Most supplements out in the marketplace are unnecessary and don't offer the powerful benefits they promise. Why? Because they've been well documented to offer nothing of value as a long-term health solution. Majority of the minerals and vitamins your body needs are typically received from a well balanced diet. Apart from what you should eat, what you shouldn't eat is also just as important in maintaining a healthily lifestyle and managing healthy acne-free skin.

Still, our diets can sometimes get off track and that's when we must consider taking necessary supplements. In this case, we're talking about magnesium. Much of the research on magnesium to date suggests it does a great job when it comes to battling the root causes of acne and most other skin disorders.

Nearly 80% of the population is deficient in magnesium. This is an astounding fact when you consider that magnesium is responsible for a over 300 biochemical reactions in the body, some of which include helping your body digest food, promote a healthy heart, and even ease migraines. The mineral also helps the body regulate blood pressure, lower the risk of cardio-vascular disease, and even reduce menstrual cramps.

How stress can cause acne and wrinkles
Have you noticed that when you're stressed, acne almost always tends to make an appearance? This is not coincidental. Under a flight or fight situation, your body releases hormones known as adrenaline and cortisol that cause us to take action in the presence of danger. Unfortunately, the type of stress we deal with on a day to day basis these days is not temporary. Work, kids, late nights, gym, all sorts of things lead to stress unless we learn to manage them. This is also the start of our long term battle against acne. This state of chronic stress leads to a build of cortisol, causing the skin to get clogged with a build-up of excreted oil also known as sebum.
Magnesium is also necessary for the enzymes that manage DNA repair and replication. Without the mineral, the skin would also be subject to a variety of harmful free radical damage and inflammation that would ultimately lead to wrinkling and skin damage.

How magnesium can help battle acne causing stress

Magnesium helps to support your adrenal system and function. Your kidneys release magnesium when cortisol is present in released in to your system, suppressing the effects of the hormone. By increasing your magnesium intake by supplementation, your body is better able to manage anxiety and stress, ultimately helping to clear your skin of acne. But make sure to self manage your stress levels because chewing a bunch of magnesium pills alone will not solve your skin woes.

Nutritionist and author of The Magnesium Miracle Carolyn Dean links the shortage of magnesium to the shortage of minerals found in our dirt, stating "Magnesium has been farmed out of the soil, so it's not in vegetables, and animals don't get it from the plants that they eat."

While we'll go over topical methods of supplementing magnesium into your routine, try to include the below foods into your diet for gaining the most benefits.

Magnesium-rich foods that are great for the skin:

- Dark green leafy vegetables such as kale, spinach, collard greens and swiss chard
- Spirulina, kelp, and seaweed
- Fish: sardines, mackerel, tuna
- Avocados
- Figs, prunes, raisins, apricots
- Bananas
- Chocolate, preferably higher cocoa content
- Nuts and seeds
- Beans and lentils
- Whole grains

While most oral magnesium products have a laxative effect and people are sensitive to that, try applying magnesium to your skin. According to the Office of Dietary Supplements at the National Institute of Health, the recommended about of magnesium is 310 to 320 milligrams per day. While topical application may not deliver the most potent amount, creams and lotions with magnesium can help.

Other benefits include improving your skin's overall appearance by inhibiting the over production of sebum that leads to acne and other disorders of the skin. To get a clear and even skin tone with the highest level of magnesium absorption, transdermal magnesium may be for you. People with redness or rosacea tend to rely on it because of its calming agent on sensitive skin.

- Sulphur

When you think of sulfur (also spelled sulphur), you might wrinkle your nose—thoughts of stinky brimstone rising to the top of your mind. However, sulfur has been used for centuries as a form of acne treatment, and this mineral is actually still used to fight acne bacteria today.

Thanks to its skin shedding and antibacterial properties, you'll likely see sulfur listed as an ingredient in many acne and skin care products.

From 200 A.D. to present day, sulfur has proven useful in many skin care regimens. However, with so many newly discovered acne fighting ingredients on the market, is it still a worthy treatment?

Sulfur and Skin Care through the Ages
Sulfur's been used in skin care regimens for thousands of years. There have been documented stories about Romans bathing in warm water filled with sulfur to help combat acne symptoms and the Ancient Egyptians used sulfur to create a salve designed to treat eczema and acne almost 5,000 years ago. Traditional Chinese Medicine also found this smelly mineral to be useful, putting it in skin care ointments over 2,000 years ago.

In the 1950s, sulfur could be found in a foam, created to apply directly to broken skin. It maintained its worldwide popularity in acne treatment products for decades. These days, sulfur can be found in three main products: sulfur soaps, sulfur ointment, and sulfur face masks—luckily, many companies have been able to reduce the mineral's telltale smell of rotten eggs in their products.

Typically used as a leave-on treatment, there are a bevy of topical sulfur products to choose from, but just how effective—and safe—is this active ingredient? First, it's important to examine why sulfur has been such a popular skin care treatment option throughout the centuries.

How Does Sulfur Help Acne?

To understand how sulfur combats an acne diagnosis, it's important to understand just how those pesky pimples come about. Hormone changes can increase the production of androgens, a male hormone that stimulates the production of sebum in the skin's sebaceous glands. When too much sebum is produced, it can combine with leftover dead skin cells and P. acnes bacteria, eventually clogging up the pore. As this mixture becomes trapped beneath the skin, inflammation occurs, resulting in an often painful acne lesion.

Sulfur works to dry out the skin; as acne is caused by the excess production of oil in the sebaceous glands, sulfur's ability to cut down on sebum can help those with acne-prone skin combat frequent breakouts.

Sulfur is also a keratolytic; when applied topically to the skin, sulfur causes the top layer of the epidermis to dry and peel off. This can help slough away dead skin cells to keep pores unblocked, allowing a fresh new layer of smooth skin to grow in its place. Sulfur also has antibacterial properties, helping to combat the presence of P. acnes bacterium that can cause infections within the dermis.

However, sulfuric acid treatments aren't right for all skin types and grades of acne. Many skin care experts believe that using sulfur on acne can help those with oily skin types, as it dries out the skin, getting rid of excess oil that can clog pores.

In contrast, sulfur can react negatively with those who have combination or dry skin types, as the excess drying effect can result in painful irritation and flaking skin. When it comes to severity, sulfur treatments for acne work best on patients suffering from mild to moderate acne; it generally can't relieve cystic breakouts or severe acne issues.

What Else Is Sulfur Used for in Skin Care?
Beyond acne treatment, sulfur is used for a host of skin ailments, including seborrheic dermatitis, rosacea, eczema, and dandruff. As sulfur assists in the shedding of excessive skin growth and fights bacteria found on the skin's surface and within the pores, it can prove to be a useful treatment option for many of these skin conditions and can often be found in varied shampoos and soaps.

- Zinc

Let's be honest. Zinc isn't the first thing that comes to mind when you think about getting beautiful skin. In fact, you've probably never even thought about it before. But this "trace mineral" (so-called because you need only a very small amount each day) is found in every single cell in your body, and more than 100 different enzymes need it to function. Skin cells are particularly dependent on zinc's powerful properties—in fact, the top layer of your skin is concentrated with up to six times more of the mineral than is found in the lower layers.

Here's how zinc keeps skin healthy and glowing from the inside:

IT WORKS AS AN ANTIOXIDANT. Though not technically an antioxidant (like vitamins C and E, for example), zinc is a key part of your skin's dietary defense squad. The mineral lessens the formation of damaging free radicals and protects skin's lipids (fats) and fibroblasts—the cells that make collagen, your skin's support structure—when skin is exposed to UV light, pollution and other skin-agers.

IT HELPS HEAL AND REJUVENATE SKIN. When you cut yourself, zinc goes to work. First, the amount of the mineral in the skin surrounding the cut increases as enzymes and proteins ramp up to protect against infection, control inflammation and produce new cells and transport them to close up the broken skin. But even healthy, intact skin relies on zinc for new cell production and the function of cell membranes.

IT MAY WARD OFF ACNE FLARE-UPS. Pimples develop when a buildup of oil, bacteria and skin cells block pores, causing the skin around the pore to turn red, swollen and tender. Zinc, which boosts immune function, may help control that inflammatory response. What's more, because zinc regulates cell production and turnover, and can help reduce the amount of natural oil your skin produces, it may prevent pores from clogging in the first place.

HOW MUCH SHOULD YOU GET?

You don't need much. Men should aim for at least 11 mg per day, while women only need 8 mg. Oysters, crab and lobster are a few top sources of zinc, but chicken, lean beef, beans, chickpeas and fortified cereals can also help you meet your daily goal. A balanced diet can often provide the zinc you need, but if you're considering a zinc supplement—or wondering if you might benefit from one—talk to your doctor.

Don't forget about zinc's topical power too. Here's how your skin benefits when you apply it to your outside:

IT'S AN EFFECTIVE SUNSCREEN. Unlike chemical sunscreen ingredients, which absorb into the top layers of your skin to filter out the damaging part of the sun's rays, zinc oxide (a zinc-containing compound) acts as a physical block, stopping UV light from penetrating your skin altogether. Although the zinc acts as a shield, the particles are micronized, so lotions containing it rub in like any other product and look invisible (you don't need to fear the tourist's swipe of white zinc oxide across the nose anymore). Even better, because zinc doesn't absorb deeply, it's less likely to irritate skin like other ingredients might—which is why it's often found in sunscreens for sensitive skin and products for kids and babies. Look for "zinc oxide" listed under active ingredients on the label.

IT CLEARS UP DANDRUFF AND RELIEVES AN ITCHY SCALP. Zinc pyrithione can be used to treat psoriasis, eczema and other skin conditions, but it's best known for diminishing dandruff. (You'll see it listed as the active ingredient on anti-dandruff shampoo and conditioner bottles.) Dandruff occurs when a common fungus that lives on the scalp grows out of control. The resulting irritation and inflammation causes skin cells to flake off and the scalp to become irritated and itchy. Zinc pyrithione not only helps control the growing bacteria, but also reduces the amount of oil that it feeds on, while its anti-inflammatory properties calm irritation and relieve itchiness. Even if you don't technically have dandruff, this form of zinc will still relieve a dry, itchy scalp. When using a zinc pyrithione product, massage it into your scalp and let it sit for a few minutes before rinsing.

IT HELPS HEAL RASHES AND OTHER SORES. A common ingredient in hemorrhoid treatments and diaper rash creams, zinc oxide helps heal skin in many of the same ways dietary zinc does: It reduces inflammation, regulates immunity and may stimulate the production of new cells. It's also an anti-microbial, which can soothe and treat skin quickly; research suggests it may shorten the duration and reduce the severity of a cold sore, for example. Scan the ingredient list for zinc oxide when looking for topical treatments to help a rash or sore.

Chapter 15

Hormones needed for anti-aging

Although the wisdom that comes with age is nice, the wrinkles, ailing joints and general frailty are not. That's why it is no surprise that 'anti-aging' is such a hot topic. Luckily, the key to anti-aging is within your reach. When it comes to relaying messages, inducing reactions and protecting tissue – it's all about your hormones.

Your hormones are chemical messengers that are keeping your body functioning. From regulating metabolism and growth to controlling immune function and reproduction, hormones are major players in all that you do and all that you are – physically at least.

Most people associate human growth hormone and DHEA with aging, but progesterone, testosterone, estrogen and cortisol play a role in aging as well. If the delicate balance of any one of these hormones is destroyed, it can take a serious toll on your body, mind and spirit – and be mistaken as classic signs of aging.

DHEA

DHEA is a precursor hormone produced from cholesterol by the adrenal glands inside the body. This hormone plays a crucial role in the formation of the sex hormones – estrogen and testosterone – as well as fuels the transformation that occurs as the body grows and matures.

Around your mid to late twenties, DHEA begins a gradual decline, which contributes to the aging process. By the age of 70, you generally have less than 10 percent of the DHEA you had in your twenties. Therapy with DHEA can be helpful in treating auto-immune disorders, obesity, dementia, osteoporosis and chronic fatigue. In some cases, it may be more effective to replace testosterone or estrogen to correct imbalances rather than DHEA , due to its role in converting to the sex hormones.

Progesterone

Progesterone serves multiple functions in the body of both men and women. It is vital for regulating the sleep cycle, as well as boosting immunity and brain function. In women, progesterone is an essential hormone of the reproductive process. Many women will experience fluctuations of progesterone throughout their life cycles. An imbalance of this hormone can lead to symptoms classically associated with aging, such as poor sleep, moody swings and foggy thinking.

Progesterone also serves as a precursor hormone, converting to estrogen, testosterone or cortisol on the steroid hormone pathway. Low levels of progesterone can lead to increased levels of cortisol and low levels of the sex hormones, which will trigger impaired immune function and a host of other issues associated with hormone imbalance.

Growth hormone

The hormone HGH (human growth hormone) is secreted by the pituitary gland and is crucial for normal development and maintenance of organs and tissues - especially in children. HGH enhances tissue growth, increases muscle mass and strengthens bone density throughout the life cycle.

Hormone therapy with HGH is most commonly used in the medical industry to treat children with stunted growth or young people suffering from hormone deficiencies. In a few cases, HGH may be necessary for treatment of hormone imbalance in adults, but this is rare.

When FDA-approved HGH is prescribed, it is in the form of a powder with a diluting solution for use as an injection. This is the only formulation that has been approved by the FDA. Of the HGH sprays and pills that are available without a prescription, few have been proven to be effective.

Testosterone

Testosterone is the principle hormone in the group of hormones known as androgens. Although classically associated with men, testosterone serves functions in both men and women.

In men, testosterone deepens the voice during puberty, stimulates the growth of facial and body hair, increases sexual drive, and is responsible for sperm production. In regards to aging, testosterone contributes to energy, memory, moods, muscle mass and strength, and sexual stamina and performance.

Women must keep testosterone levels in fine balance, as it plays an integral role in mood, energy, weight and sex drive. Without enough testosterone, gaining and sustaining lean body mass can be difficult. As testosterone declines with age, it impacts weight management and sex drive.

Estrogen

The term estrogen collectively refers to the hormones, estradiol, estrone and estriol. Each has a different chemical structure and function. Like testosterone, estrogen plays a role in the physiology of both men and women. However, estrogen plays a greater role in the aging process for women than in men.

The primary role of estrogen in the female body is to stimulate growth and development of sexual characteristics and reproduction, induce the changes of the breasts during adolescence and pregnancy and aid the growth of the uterine lining during the follicular phase of the menstrual cycle. Estrogen is responsible for hundreds of functions in the female body including protecting women from heart disease and colon cancer. It also regulates several metabolic processes, including cholesterol levels and bone growth.

In men, estrogen aids in the facilitation of many of the physical changes men experience during puberty, such as chest and facial hair growth, muscle development and deepening of the voice. Estrogen also helps to protect the bones and the brain in men.

Pursuing hormone therapy for anti-aging

A specialized physician can help you determine if hormone therapy will treat or reverse your hormone imbalance and the symptoms often linked to advanced aging. Lab testing, followed by a thorough consultation, will ensure you get the quality treatment you need to live better and longer.

Chapter 16
Advanced Anti-Aging Techniques
Growing old is just a natural part of life, and it can't be avoided. At least, that's what most of us have accepted when it comes to aging. But for decades, scientists have been working to unlock the key to keeping us young and healthy.

Though it might sound outlandish, there have been some incredible breakthroughs in the field of anti-aging technology. As we continue to learn more about the human body and how it functions, we grow ever closer to making old age a thing of the past. Here are just some of the startling innovations that could make reverse aging a reality.

1. Stem Cell Technology: Reprogramming Aging Cells
In 2016, researchers from Salk Institute for Biological Studies in California succeeded in reprogramming the cells of aging mice using induced pluripotent stem cells. These stem cells are generated from adult cells, and allowed scientists to reprogram skin cells to an embryonic state.

In their study, researchers found that the mice whose cells had been reprogrammed lived 30% longer than the non-reprogrammed control group.

There are risks involved with using induced pluripotent stem cells, and human trials are definitely a long way away, but the study provided some of the most interesting data so far on the use of stem cells in anti-aging efforts.

2. Targetting Mutant mtDNA: Repairing Aging Cells

In November of 2016, researchers at CalTech and UCLA discovered a method of manipulating the mitochondria of cells to effectively repair DNA. Cells typically contain two types of DNA found in the mitochondria - normal mtDNA and mutant mtDNA. The study noted that the build-up of mutant mtDNA over time causes cells to age and eventually die.

The study was concerned with whether or not autophagy, or the process through which cells eat themselves, could be used to target the mutant mtDNA and prevent the aging process.
They were successful in increasing mitophagy activity in a fruit fly, noting a distinct reduction of mutant mtDNA in its muscle cells. Time will tell if similar techniques can be used to reduce aging or repair aged cells in humans.

3. Activating Splicing Factors: Crafting Reversalogues to Encourage Cell Division

In late 2017, BMC Cell Biology published research from the University of Exeter and the University of Brighton, both in the United Kingdom, which presented a breakthrough in reverse aging technology. The work built upon a previous University of Exeter discovery, which found that splicing factors in cells become inactive with age.

The study found that by introducing reversalogues, similar to the chemical resveratrol which is found in red wine, splicing factors could be reactivated in older cells. This meant that the cells would continue to divide like younger cells, rejuvenating the cells and preventing cell aging and death. The implications of this research could mean longer life, reduced signs of aging, and better health in later years.

4. Rejuvenate Bio: Reversing the Process of Aging in Dogs

Who wouldn't want a puppy that stays cute and young forever? A secretive start-up within Harvard, known was Rejuvenate Bio, has been working on technology to reverse the process of aging in dogs. Though little is known of the company's research methods, it's been suggested that they're modifying certain genes as a way of targeting and eliminating risks of heart disease, kidney disease, and more.

The company is focusing primarily on Cocker Spaniels and Doberman Pinschers for now, as both breeds have short lifespans due to their genetics. Rejuvenate Bio feels that their reverse aging on dogs is a marketable service, which will hasten FDA approval and pave the way for human trials to follow.

5. Senolytic Drugs: Combining Pre-Existing Medications to Achieve Reverse Aging

Just this month Nature Medicine published research from the National Institute of Aging which found that combining and injecting pre-existing drugs could extend lifespan and delay age-related health issues. The study was conducted on mice, who were treated with dasatinib and quercetin. Dasatinib is usually used to treat forms of leukemia, whereas quercetin is found in fruits and vegetables.

When treated with the medications, researchers found that senescent cells were selectively killed and the deterioration of speed and mobility of the mice significantly decreased.

In fact, naturally aging mice treated with the drugs in the study lived 36% longer than their control group counterparts. If similar results can be replicated in humans, it would mean that we already have drugs that could reverse aging and lengthen our lifespans.

6. Synthetic Peptides: Intervening in the Aging Process

Last month saw new research emerge from the Marshall University Joan C. Edwards School of Medicine, which suggested that the Na/K-ATPase oxidant amplification loop (NAKL) could potentially be targetted for anti-aging interventions. The researchers were also successful in testing pNaKtide, a synthetic peptide that could further reduce risks of illness and the effects of aging.

The researchers first tested their theories on mice who were raised on a Western diet, to antagonize the NAKL. This provided the evidence of aging, which was then treated with pNaKtide.

The treatment slowed the aging process that had been spurred on by the NAKL. They then tested the treatment on human dermal fibroblasts, and recorded similar results. This would suggest that if we can effectively target the NAKL and intervene with a treatment of pNaKtide, we could potentially slow the aging process significantly.

7. Smoothing Cells: Using Viruses to Smoothen Cell Wrinkles
It's not just your skin that wrinkles as you age, turns out our cells get wrinkly too. In May of this year, researchers from the University of Virginia School of Medicine found that many effects of aging, such as fatty liver disease, could be the result of cell nuclei wrinkling. When the nuclei wrinkle, it prevents DNA from functioning as it should.

So what do you do if you have wrinkly nuclei? Well, according to their research, we could be able to utilize viruses to smooth out the nuclear membranes. They believe that viruses could be modified to carry and deliver lamin - a protein that can smooth the cells. Through doing this, the cells could function like young cells again, reversing the effects of aging and protecting against multiple health risks.

8. Young Blood: Pumping Youth Back Into Veins

It might sound like something out of a vampire story, but young blood might actually be the key to reversing the aging process. In February of this year, Cell Reports published a study that suggested that the blood of young people could be instrumental in fighting against aging.

In the study, researchers found that when the blood from younger mice was introduced into the blood of older mice, it stimulated neuron and stem cell production in the older mouse's brain. This effectively led to their brains functioning more like those of their younger counterparts, reversing the effects of aging on their cognition.

Clinical trials are already underway where plasma from 16 to 24-year-olds will be introduced into the systems of elderly humans. So far results have been promising, but it will be some time before everyone starts loading up on young blood to stay youthful.

9. Anti-Aging Pills: Treating Age With Medication Researchers from the Paul F. Glenn Center for the Biology of Aging at Harvard Medical School believe that they could be able to create anti-aging pills. This comes as a result of a study published in March 2017, which saw scientists use nicotinamide adenine dinucleotide (NAD+) to reverse the signs of aging in mice.

NAD+ is a molecule found in cells which is crucial for maintaining healthy cell function and regulating cellular aging. In the experiment, the scientists put a few drops of NAD+ into the mice's water. Within hours, levels of NAD+ in their cells had risen significantly. Within a week, the tissues of the NAD+-fed mice had reversed to youthful levels.

A human trial was conducted in November 2017, which showed similarly promising results. If NAD+ supplements can get FDA approval, we might have a reverse aging pill on our shelves soon.

10. Reverse Aging With Cannabis: Improving Brain Function With THC

Strange as it may sound, cannabis might actually be the key to reversing the signs of aging and improving cognitive ability. That's what was suggested by research from the University of Bonn and The Hebrew University of Jerusalem which was published in May of last year.

The study saw researchers successfully reverse the biological state of the brain in mice aged 12 months and 18 months. The test mice were treated with THC in small, non-intoxicating doses for a period of four weeks. They were noted as out-performing the control group who had been given a placebo, and were shown to have similar cognitive performance to younger mice. This could mean that non-intoxicating THC treatments might allow older humans to regain youthful levels of cognition.

11. Anti-Aging Bacteria: Using Bacteria to Create Anti-Aging Pills

Some researchers believe that the key to reversing the aging process lies in a bacterium found in Easter Island. Rapamycin is already used in transplant medicines and as an immunosuppressant, but scientists now believe that it could be used to reverse the effects of aging.

In a series of experiments, rapamycin was proven to postpone death in worms, flies, and mice. More recent studies are focusing on dogs as test subjects, with an aim to someday use rapamycin in an anti-aging pill for humans.

Though many companies are scrambling to be the first to have a rapamycin pill approved and marketed, it will be some time before they hit shelves for human consumption.

12. Gene Deletion: Deleting Selective Genes to Increase Lifespan

In 2015, researchers from the Buck Institute for Research on Ageing and the University of Washington revealed that, after ten years of research, they had succeeded in identifying and delcting genes that could prolong life. The researchers identified 238 cells which, when removed, resulted in a 60% increase in lifespan.

Many of the genes identified are also present in mammals, meaning that theoretically the same process could be applied to humans. That said, it took 10 years for researchers to find and delete the genes responsible for aging in yeast through a process of trial and error. Replicating the results in humans could take a very long time, though the results are still encouraging.

www.ingramcontent.com/pod-product-compliance
Lightning Source LLC
Chambersburg PA
CBHW030655220526
45463CB00005B/1793